Praise for *Lean Retail and Wholesale*

"Paul Myerson's book is essential reading for both students and professionals alike who want to learn how Lean provides a competitive edge in today's fast-paced, multichannel, and cost-conscious environment. Myerson simplifies complex topics, providing guidance on executing Lean across the extended supply chain including manufacturing, distribution, retail, e-tail, and the home office. The author adds depth to each topic by referencing examples from his breadth of consulting experience. In the end, Myerson understands that Lean's positive impact multiplies with its wide and collaborative application across the extended supply chain."

Mark Temkin
Director, Demand Planning
Barnes & Noble Inc. / Nook Media LLC

"This book provides an enlightening perspective on the application of Lean principles to the increasingly challenging worlds of the retail and wholesale sectors. Considering the accelerating pace of change in these important areas of the global economy, including the contemporary relevance of omnichannel business operations, the ability to apply Lean principles will increasingly help to differentiate those who are successful from those who are not."

C. John Langley Jr., Ph.D.
Professor of Supply Chain Management
Penn State University

"As the Lean philosophy is relatively new to the retail and wholesale world, this book does a great job of applying Lean concepts, tools, and real-world examples to this part of the supply chain. This book will be a great resource for anyone who is responsible for driving continuous supply chain improvement at their organization. It should be on every manager and executive's bookshelf."

Robert J. Trent, Ph.D.
Supply Chain Management Program Director
Lehigh University

About the Author

Paul A. Myerson has been a successful change catalyst for a variety of clients and organizations of all sizes. He has more than 30 years of experience in supply chain strategies, systems, and operations that have resulted in bottom-line improvements for companies such as General Electric, Unilever, and Church and Dwight (Arm & Hammer). Mr. Myerson holds an MBA in Physical Distribution from Temple University and a BS in Logistics from The Pennsylvania State University. He is currently Professor of Practice in Supply Chain Management at Lehigh University and Managing Partner at Logistics Planning Associates, LLC, a supply chain planning software and consulting business (www.psiplanner.com). In the past, Mr. Myerson has served as an adjunct professor at several universities, including Kean University and New Jersey City University. He is the author of *Lean Supply Chain and Logistics Management* and has published a Lean supply chain and logistics management simulation training game. Mr. Myerson currently writes a column on Lean supply chain for *Inbound Logistics Magazine* and a blog on the same subject for *Industry Week* magazine.

Lean Retail and Wholesale

Use Lean to Survive (and Thrive!) in the New Global Economy with Its Higher Operating Expenses, Increased Competition, and Diminished Consumer Loyalty

Paul A. Myerson

New York Chicago San Francisco
Athens London Madrid
Mexico City Milan New Delhi
Singapore Sydney Toronto

Cataloging-in-Publication Data is on file with the Library of Congress.

McGraw-Hill Education books are available at special quantity discounts to use as premiums and sales promotions, or for use in corporate training programs. To contact a representative please visit the Contact Us page at www.mhprofessional.com.

Lean Retail and Wholesale: Use Lean to Survive (and Thrive!) in the New Global Economy with Its Higher Operating Expenses, Increased Competition, and Diminished Consumer Loyalty

1 2 3 4 5 6 7 8 9 0 DOC/DOC 1 9 8 7 6 5 4

ISBN 978-0-07-182985-4
MHID 0-07-182985-7

The pages within this book were printed on acid-free paper.

Sponsoring Editor
Judy Bass

Acquisition Coordinator
Amy Stonebraker

Editorial Supervisor
David E. Fogarty

Project Manager
Shruti Chopra, MPS Limited

Copy Editor
Alice Manning

Proofreader
Nisha Rawat, MPS Limited

Indexer
Judy Davis

Production Supervisor
Pamela A. Pelton

Composition
MPS Limited

Art Director, Cover
Jeff Weeks

This book is dedicated to my family and to those who strive to continuously improve themselves and the world around them

CONTENTS

PART III

Lean Forward

PREFACE

We live in exciting, yet challenging times.

During the past 20 years or so, we have seen a shift in the balance of power from manufacturers to giant retailers such as Walmart, Target, and Home Depot.

More recently, these same retailers and wholesalers have had to face a challenging economy following the Great Recession, which has made consumers more value-focused than ever before.

Multichannel marketing models are now commonly used to reach the consumer using a combination of indirect and direct communication channels, including traditional "brick-and-mortar" stores, Web sites, mail order catalogs, direct mail, e-mail, and mobile, among others.

Internet-based models such as e-commerce and m-commerce have enabled consumers to have better information and more choices, thus increasing their buying power. Multichannel marketing not only helps retailers and wholesalers, but also opens the possibility that manufacturers can communicate directly with consumers, giving them an opportunity to shift the balance of power back in their favor to some degree.

To help them cope with this complex and competitive global economy, retailers and wholesalers of all sizes now have access to various technologies to improve visibility, collaboration, and efficiency up and down the supply chain.

However, technology can only enable a good process. Retailers and wholesalers need to harness the power of their employees using a team-based, customer-focused continuous improvement tool such as Lean to identify and eliminate waste in their organization, wherever it may be.

While a limited number of large retailers have pushed manufacturers to become leaner and more efficient through programs such as Quick Response (QR) and Efficient Consumer Response (ECR), most have not created in their organizations the type of Lean culture and philosophy that is necessary for long-term success.

As a result of the limited application of Lean in retail and wholesale, there isn't a lot that has been written on this subject to help retail and wholesale professionals find their way. The closest fit to retail and wholesale would be books written on Lean in the service and apparel industries—not a perfect fit for retail and wholesale.

The purpose of this book is to give practitioners the necessary tools, methodology, and "real-world" best practice examples to enable them to implement Lean successfully in their retail and wholesale organizations.

This book is organized so that the reader can first gain some perspective on how retail, wholesale, and Lean have evolved in recent years. This is followed by explanations and examples of both basic and advanced Lean tools, along with specific implementation opportunities in the retail and wholesale industries. A Lean implementation methodology with critical success factors is then identified and described.

In addition to real-world best practice examples throughout the book, there are also several case studies on the subject in Appendix A that provide insight into how others have successfully implemented Lean in their organizations.

Additionally, this book comes with a downloadable Lean Opportunity Assessment tool and training slides to help you make the transition from understanding Lean to implementing it in your company or your client's company. (Go to www.mhprofessional.com/leanretail.)

Paul A. Myerson

PART I

Current State

CHAPTER 1

Introduction: Using Lean as a Tool to Survive and Thrive in Retail and Wholesale

Lean, a team-based form of continuous improvement that focuses on the identification and elimination of non-value-added activities, or "waste" from the viewpoint of the customer, has historically been thought of largely as a tool for manufacturers.

The reality is that Lean thinking in a variety of manufacturing sectors, such as consumer goods, apparel, and food and beverage, has actually been accelerated by a select group of forward-thinking retailers such as Walmart and OfficeMax.

These retailers have dramatically changed the way products are ordered, moving inventory rapidly through their distribution centers to the stores by utilizing sales data gathered electronically at checkout, sharing those data with suppliers, and primarily using bar codes to manage and accelerate the flow of product from the manufacturer to the store shelves.

Some leading-edge large retailers have placed an emphasis on just-in-time (JIT) delivery to reduce inventory costs and increase in-stock levels. They have made significant investments in technology to allow stores to share point of sale (POS) data with their suppliers, and in some cases they have allowed manufacturers and wholesalers to replenish inventory for the retailer automatically. Some Lean retail software systems can automatically place a new order for an item as soon as that item is scanned at the checkout counter.

At least partially as a result of these initiatives, manufacturers have been forced to respond to orders more rapidly, using collaborative approaches to distribution, forecasting, planning, production, and even supplier relations.

Yet, although this revolution has been going on in manufacturing and for a select few large retailers, the majority of retailers and wholesalers have implemented only some, if any, Lean concepts, and in a piecemeal fashion. Most of those activities have focused largely on their suppliers upstream, rather than on identifying what does or doesn't add value to their customers in the first place.

In this book, we will demonstrate that through a greater use of Lean concepts and tools, retailers and wholesalers can utilize a comprehensive, holistic Lean methodology throughout their businesses to reduce costs and improve service and profitability, even while they are faced with shorter product life cycles and constantly changing consumer tastes, resulting in a demand for mass customization and an explosion in the number of items carried.

Why Implement Lean in Retail and Wholesale?

Retail businesses allow manufacturers of goods to reach more customers than they could reach if they sold directly, resulting in greater revenues, as well as achieving cost efficiencies, improving cash flow, and focusing on their key competencies.

Wholesalers also perform a similar (intermediary) role, as they are businesses that buy large quantities of goods from various producers or vendors, then warehouse those goods and resell them in smaller lots to retailers and to industrial, commercial, institutional, or other professional business users.

As a result, retail and wholesale make up a major portion of the U.S. and world economy. According to the U.S. Department of Commerce, retail sales are almost $5 trillion (supplied both by manufacturers directly and through wholesalers), which is approximately one-third of the total economy.

According to the Bureau of Labor Statistics, retailing employs about 15 million people, while wholesale trade employs another 5.7 million people. This doesn't include all of the related professions that support and supply these sectors [Bureau of Labor Statistics, 2013].

By taking a systematic and holistic approach to the identification and elimination of non-value-added activities throughout their businesses, retailers and wholesalers can improve their margins and throughput.

Needless to say, the potential improvements we can attain in these sectors can have massive implications for our economy as a whole.

Challenges in Retail and Wholesale

During the "Great Recession," which lasted from December 2007 through June 2009, manufacturing, wholesale, and brick-and-mortar retail sales took a heavy beating. Even by the fourth quarter of 2010, they still had not fully recovered. Many people are predicting a "new reality" of slower growth for years to come. There also seems to be a change in the spending habits of the Generation X and Y population, similar to the penny-pinching effect on those who endured the Depression in the 1930s. In fact, "according to the AICPA and Ad Counsel, 94 percent of 25 to 34 year olds want to save for the future, but 4 in 10 have difficulty putting aside just $25 a week" [Murad, 2012].

This new reality, combined with an ever-expanding list of offerings, shorter lead times and product life cycles, and the growth of retail e-commerce, may force retailers to operate with fewer employees and lower operating costs for years to come.

As we enter this slow-growth recovery from the recession, there may be some lessons learned for operations management, including the following:

- The Great Recession clearly focused tremendous managerial attention on managing and improving processes. . . . Unfortunately, even The Great Recession failed to convert most managers and organizations into process zealots, and now as in the past, the majority still only half-heartedly and inconsistently improves processes. . . . How do we evolve process improvement so that it stays effective but is more easily deployable in all kinds of contexts?
- Workforces in many organizations were pared during The Great Recession with little intention of ever replacing those employees, and the reality going forward is that organizations will be looking for ways to increase productivity given reduced headcounts.
- A punishing lesson of The Great Recession was that supplying customers with near limitless variety also courted financial disaster. . . . The way that customer demand so quickly plummeted at the start of The Great Recession, and then so

stubbornly refused to recover, suggests that most manufacturers and retailers are not going to resume offering near limitless variety anytime soon.

- Strong operations make the biggest profits in good times and help ensure survival in bad. We have known the first part of that sentence to be true for decades, but the second half was only a theory up until recent events so thoroughly tested it. Now we know the entire sentence is true and organizations with the strongest operations prevail in both good times as well as bad, which is why companies like . . . Southwest Airlines, Wal-Mart, and Zara easily survived The Great Recession while so many others floundered in its wake [Frohlich, 2013].

Need for Lean in Retail and Wholesale

All this points to the need for a retail and wholesale organizational culture that supports a Lean thinking philosophy with a consistent, simple methodology for process improvement in order to survive and thrive in the coming years. Lean thinking can make this happen if it is implemented correctly.

Lean is much more than reducing inventory and trimming your workforce in retail and wholesale. It's about a philosophy of continuous improvement that focuses on identifying and eliminating activities that are non-value-added from the viewpoint of the customer. Our employees don't necessarily have to work harder; instead, they have to work smarter.

This book will examine Lean opportunities in retail and wholesale from the viewpoint of retail strategy, merchandise management, and store and distribution operations.

Retail Strategy Viewpoint (Including Sales and Marketing, Location, Human Resources Management, IT, Supply Chain Management, and Customer Relationship Management)

A retail strategy is all about committing the company's resources to developing long-term advantages relative to the competition.

In a strategic sense, to some degree, power has shifted from manufacturers to Lean retailers, such as Walmart, Federated Department Stores (Macy's, Bloomingdale's), the Gap, Sears, and Home Depot. These powerful retailers now insist on low prices and refuse to carry excess inventory, expecting manufacturers to provide rapid and frequent replenishment of retail products based on real-time sales.

Those organizations with less power—small- to medium-sized retailers and wholesalers—can also move in this direction, but they must be more selective about it, as they have limited resources. We will explore ways to do this throughout this book.

All businesses, including retailers and wholesalers, should have an overall mission statement (i.e., a statement of the purpose of the organization and its reason

for being), along with departmental or functional missions. Strategies are then required to achieve the overall and functional missions.

Traditionally, organizations have used tools such as SWOT analysis (strengths, weaknesses, opportunities, and threats) to develop a road map for success by exploiting strengths and opportunities, neutralizing threats, and avoiding weaknesses to develop successful strategies. This can and should include a Lean strategy to be used for a competitive advantage.

For Lean to be successful in a retail or wholesale organization, departmental strategies must be aligned with and supportive of a companywide Lean strategy throughout the business.

Sales and Marketing

We need to look past just improving internal transactional activities like processing orders and creating ad copy to include integrating customer and companywide value streams, integrating market knowledge, and creating actual "customer pull." The results are not just better sales and marketing processes, but greater overall achievement of an organization's strategy.

Location

The location decisions for retail and wholesale/distribution firms are critical and complex strategic decisions that cannot be taken lightly. Therefore, the decision-making process should be examined not just for efficiency, but for effectiveness. The location of a distribution center can have a significant impact on both cost and service, and we all know the saying that "location, location, location" is critical when it comes to retail locations in order to maximize revenue.

Human Resource Management

Lean requires a different way of looking at your employees and how they look at themselves. It requires a commitment from top management, sometimes requiring a cultural change. More is asked of employees, but the rewards are plenty, including increased morale, greater productivity, and career growth.

Information Technology

Just walking through a store, one can quickly see the impact that technology has had on productivity in recent years. Technologies ranging from self-checkout to automated price checking have changed the way we shop. Bar codes and radio frequency identification (RFID) technologies have changed and enhanced the interactions between retailers, wholesalers, and manufacturers. Technology must be able to adequately enable and support a Lean (or any operational) strategy.

Supply Chain and Logistics Management

Companies may have more than one supply chain, depending on the business they are in. In all cases, the supply chain and logistics organization is the "glue" that holds everything together. So it is critical that information and product flow efficiently both upstream toward suppliers and downstream toward customers.

Customer Relationship Management

A key to efficiency and waste reduction (which we'll define and discuss later in the book) is having an integrated customer relationship management (CRM) process. This ties a variety of activities together so that all of them are focused on adding value to the customer.

Ultimately, an organization will have an overall operational strategy that might target a combination of characteristics, such as quick response, low cost, product differentiation, flexibility, high quality, superior performance, and the like. The functional strategies must then follow to support the operational strategy, and that's where Lean fits in as the ideal philosophy to ensure value to the customer with minimal waste.

In this book, we will explore how the various functional strategies can and must be aligned and targeted to ensure a Lean, efficient business that is focused on adding value to the customer.

Merchandise Management Viewpoint (Planning, Buying, Pricing, and Communications)

Merchandise management is the process of developing, securing, pricing, supporting, and communicating the retailer's merchandise offering. Ultimately, it means having the right product at the right price and the right time.

By not executing this in a Lean and efficient manner, a great deal of waste is created that doesn't add value to anyone.

Merchandise management (and store and distribution operations) is more tactical in nature and is about implementing the retail strategy. These decisions are more short-term in nature and more about efficiency.

Inventory

From an inventory standpoint, it is important to note that manufacturers, wholesalers/distributors, and retailers have different measures of exposure to investments in inventory. If a business doesn't have the optimal inventory assortment, it can suffer from lost sales and poor customer satisfaction or overstocks, resulting in increased costs and lower profitability.

Dimensions of Inventory

There are dimensions of inventory that differ among manufacturing, wholesale, and retail. They are time duration and depth and width of commitment.

A manufacturer's inventory (raw materials, work in process, and finished goods) exposure is usually on the narrow side because it has fairly limited offerings, but deep and of long duration because of its extensive channels of distribution and relatively few manufacturing sites.

A wholesaler's inventory exposure is wider than a manufacturer's and somewhat deep, as the wholesaler supplies a variety of customers and industries. Its inventory duration is medium, as it typically buys in large quantities and sells smaller quantities to retailers. In some ways, wholesalers have the greatest risk of the three, as they may be required to hold a high variety of items with some depth, as mentioned earlier, and this can be exacerbated as a result of customer seasonality.

A retailer's exposure is wide, but not very deep. Duration is usually short except for specialty retailers. However, as mentioned earlier, some major retailers have pushed some of the responsibility for width and depth back to wholesalers and manufacturers [Bowersox et al., 2013].

Store and Distribution Operations Viewpoint

Perhaps the greatest area of waste, and therefore the greatest opportunity to apply Lean principles, is in the area of store and distribution operations.

In distribution, it's all about optimizing the trade-offs between handling costs and costs associated with warehouse space and maximizing the total "cube" of the warehouse—utilizing its full volume while maintaining low material handling costs and minimizing "travel time."

When we think of store operations and process improvement, it is helpful to look at them from the customers' viewpoint as they make their way through the store and break it into the following steps (see Fig. 1.1).

Enter

When the customer enters the store, seeking a particular merchandise, it's important that they understand immediately how to get to the area where the items they are hoping to find can be located. . . . It is sometimes difficult to clearly understand where to go. This leads to delays, frustration, and misdirection—types of wasteful activities from the customer's perspective.

Seek

Store layouts need to be effective for the customer's purpose. Principles can be applied to floor plans and workspaces to ensure that the customer can "flow" through the store in the fastest and easiest manner.

Figure 1.1 "Five points of transition" in Retail.

Find

When customers need help in finding products or information, it's important to have an appropriate staffing level to reduce wait times. Lean principles can help retailers adapt their staffing levels to accommodate peak traffic times, allowing your employees to give customers a greater level of service.

Select

After the customers find the products or information they need, the next step is selection from the options that are available. By tracking buying patterns and the impact of in-store product placement on sales, the retailer can more appropriately tailor inventory levels.

Transact

After the customer finds the products, the payment process should be flexible, fast, and easy to navigate. Lean principles can help decrease wait times through both effective staffing and removing traffic bottlenecks. Improvements here may include incorporating better technology, such as a robust POS program to enhance service, offer loyalty program services, and find opportunities for sales suggestions and better service [Guidon Performance Solutions, 2013].

The SCOR Model and Lean Retail and Wholesale

Many Lean opportunities in retail and wholesale occur in operations, especially in the supply chain and logistics function. A strategic way to view an organization and its supply chain that was created by the Supply Chain Council is commonly referred to as the *SCOR model*. This model can be a useful tool when attempting to identify Lean opportunities in any organization.

> "The Supply-Chain Operations Reference-model (SCOR) captures the Council's consensus view of supply chain management. The SCOR model provides a unique framework that links business process, metrics, best practices and technology features into a unified structure to support communication among supply chain partners and to improve the effectiveness of supply chain management and related supply chain improvement activities. . . . The SCOR model has been developed to describe the business activities associated with all phases of satisfying a customer's demand. The Model itself contains several sections and is organized around the five primary management processes of Plan, Source, Make, Deliver, and Return." [*See Fig. 1.2; Supply Chain Council, 2013*]

The SCOR model has four levels. Level 1, as shown in Fig. 1.2, is where competition targets are set. In Level 2, companies implement their operations strategy according to how they are configured. In Level 3, they fine-tune their strategy, and Level 4 is where they implement their supply chain practices.

Version 6.0 of the SCOR model modified what was meant by "Deliver" in Level 2 to include the delivery of retail product.

In today's environment, retailers need to determine how to deliver more for less. So in terms of the SCOR model, they need to focus on improving delivery reliability, flexibility, and responsiveness by planning, sourcing or making (i.e., private label with contract manufacturers), delivering, and returning more effectively and efficiently.

For the purposes of this book, I think it's best to use the SCOR model a bit differently by grouping the retail and, to some degree, wholesale, supply chain and viewing it as Plan (retail strategy), Source or Make (merchandise management), Deliver (warehouse and store operations), and Return (reverse logistics). This will

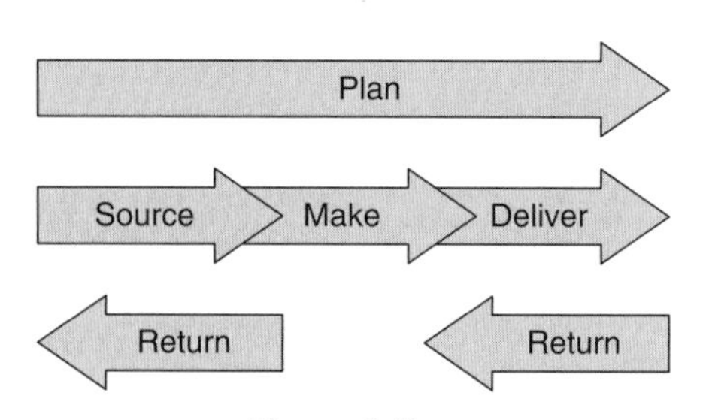

Figure 1.2

prove to be useful at later points in the book, especially when discussing performance measurements.

Making Change Successful

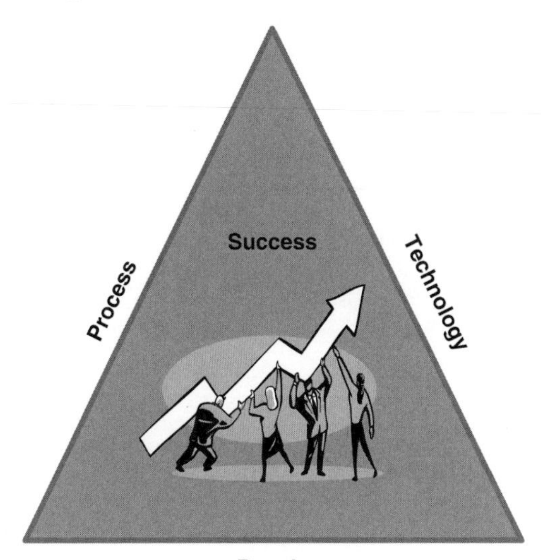

In many cases, transformation efforts focus on process improvement only, while ignoring the people (and in some cases technological) aspect of the change initiative. As a result, many of these transformation initiatives don't achieve their desired results.

It's been said that up to 75 percent of Lean and other process improvement and reengineering efforts don't reach their objectives and, as a result, don't continue over the long term.

One of the most common reasons for the failure of such initiatives is a lack of focus on the organization's culture. Any type of change initiative should focus on developing key organizational competencies around organizational culture transformation and process improvement, which can result in a more effective and sustainable change effort.

In fact, according to an IBM Global CEO study [Jørgensen et al., 2008], most of the critical challenges in successful change involve people, process technology, and leadership.

We will discuss this later in the book when we address how to sustain improvements, but in general, it's always a good idea to look at change from a combination of people, process, and technology standpoints and to make sure that they are all in harmony.

To know where we need to go, it's always a good idea to see where we are and how we got there, which is the topic of the next chapter.

CHAPTER 2

(R)evolution of Retail: From the General Store to E-commerce

All anyone has to do to see how retail has changed over the years is to walk into a neighborhood supermarket, where there will be a vast assortment of products, giving the consumer a seemingly endless number of choices and an array of technology to make the selection and checkout process smooth and easy.

Large wholesalers have similarly increased the number of choices that they offer and their use of technology to speed the distribution process, as I saw not too long ago at a large distribution center (DC) of W. W. Grainger, a Fortune 500 industrial supply company. This DC is modern and automated and can process thousands of orders per day, the majority of which are accurate and complete. Its catalog lists more than 400,000 items, and customers can purchase more than 900,000 items online.

We need to keep in mind that the journey isn't nearly complete in terms of people, process, or technology. In many cases, large retailers and wholesalers are further ahead as a result of having greater resources, but even they can find room for improvement.

While we don't want to drive forward by looking in the rearview mirror, it sometimes helps to look back before looking ahead.

First, let's get a few definitions out of the way.

Retail Versus Wholesale

As we know, retail is a critical part of a distribution channel (Fig. 2.1), which typically consists of a manufacturer, a distributor or wholesaler, and a retailer.

Retail and Wholesale Defined

A retailer purchases goods or products in large quantities, either directly from manufacturers or through a wholesaler, and then sells smaller quantities to consumers for profit. Retailing can be done either in fixed locations like stores, markets, and door-to-door, or by means of delivery (from a catalog or website).

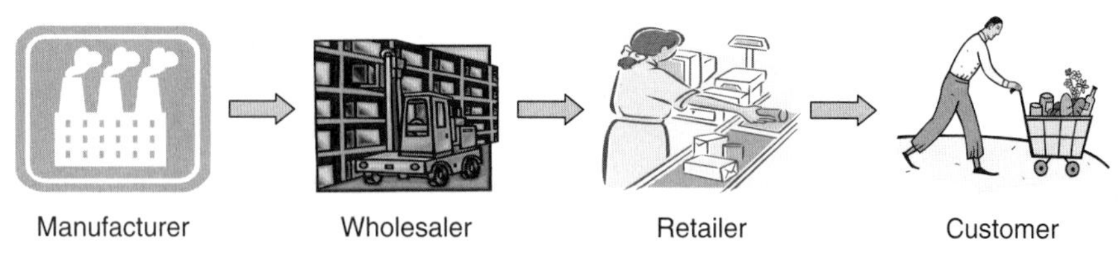

Figure 2.1 Channels of distribution.

A wholesaler, on the other hand, is a business that buys large quantities of goods from various producers or vendors, stores them, and resells them to retailers. Wholesalers who carry only noncompeting goods or lines are called distributors.

Retail Classifications and Types

The typical classifications of retail are by the type of product, and the major classes are:

- *Food products*—This includes various food and beverage products.
- *Hard goods or durable goods*—This includes appliances, electronics, furniture, sporting goods, and so on. Goods of this type don't wear out quickly and provide utility over time.
- *Soft goods or consumables*—This includes clothing, apparel, and other fabrics. These goods are consumed after one use or are used for have a limited period of time.

There are many types of retailers by marketing strategy. The major ones by product line are:

- *Department stores*—Large stores offering a wide assortment of soft and hard goods. Stores of this type typically carry a variety of categories and have a broad assortment of goods. This category includes mass merchandisers such as Walmart and Target and the so-called category killers, such as Home Depot and Bed Bath & Beyond, that dominate one area of merchandise.
- *Supermarkets*—Self-service stores that sell a wide variety of items such as food, toiletries, household products, and so on.
- *Specialty stores*—Smaller stores that focus on a particular category and provide a high level of service to their customers.
- *Discount stores*—Stores that offer a wide array of products and services, but compete mainly on price and offer an extensive assortment of merchandise at low prices.
- *Convenience stores*—Small self-service stores that provide limited amounts of merchandise (food and nonfood) at higher than average prices with fast checkout.
- *Hypermarkets*—Very large self-service stores that provide variety and huge volumes of exclusive merchandise at low margins. The operating cost is comparatively less than that of other retail formats. Hypermarkets typically combine a supermarket with a department store (Walmart and Target, e.g., have a growing number of stores with these characteristics).

- *Warehouse stores*—Warehouses that offer low-cost, often high-quantity goods on pallets or shelves. Warehouse "clubs" usually charge a membership fee.
- *E-tailers*—Stores that sell goods on the Internet. Customers can shop and order through the Internet, and the merchandise is delivered to the customers' homes.
- *Multichannel*—This is the merging of retail operations to enable customers to purchase through many different channels. These channels can include retail ("brick-and-mortar") stores, online stores, mobile stores, mobile app stores, telephone sales, and any other method of selling to a customer. The transactions can include browsing, buying, and returning as well as pre- and postsale service.

History of Retail

As mentioned previously, it's sometimes best to see where we've been before we decide where we want (or need) to go. So let's look at a brief history of retail.

Pre-World War II

Prior to 1945, retail was primarily made up of "mom-and-pop" stores and general stores. The mom-and-pop stores were family-owned and included grocery and hardware stores. General stores offered a wide variety of items.

Retail Growth (1945–1975)

This was the era of chain stores such as Sears, J. C. Penney, and Macy's. They expanded from cities into the suburbs at large malls and shopping centers. This was also a period of rapid growth for large supermarket chains such as A&P, Safeway, and Kroger.

Big-Box Stores and Category Killers (1975–1990)

During this period, there was rapid growth of mass merchandisers, such as Walmart, Kmart, Sears, and the like. It was also the beginning of what became to known as category killers: superstores that specialized in one category, such as Best Buy, Staples, and Bed Bath & Beyond.

Retail Consolidation (1990–2000)

In this period, there was much industry consolidation as the larger chains, such as Walmart, Kohl's, and Home Depot, grew bigger and smaller chains and mom-and-pop stores closed. The list of now-defunct retailers seems endless, including such names as Caldor, Ames, E. J. Korvette, and Woolworth.

There was also the start and growth of supercenters and warehouse stores that featured "one-stop shopping." This was part of Walmart's successful growth strategy.

As a result, many manufacturers had a shrinking list of retail customers that now had greater negotiating power concerning price and the supply chain.

The Twenty-First Century (2000–Present)

Consolidation has had a major effect on merchandising strategies for both shoppers and suppliers. As retailers increase the use of private-label products, there is less leverage available to major brand names. This also gives customers the opportunity to receive the same or similar quality at a lower price.

Retail e-commerce has grown to more than $200 billion per year, yet it is still only 6 to 7 percent of total retail sales. However, it is continuing to grow at a rapid pace. This, to some degree, fills the void left by retail consolidation.

In addition to pure e-tailers, brick-and-mortar retailers such as Walmart and Target have a substantial presence on the Internet with online stores (i.e., multichannel).

Generation Y shoppers have grown up with technology and are sustaining the Internet's growth as a retail channel.

Some mid-priced stores, such as Sears and J. C. Penney, have struggled, while discounters, specialty stores, and luxury stores have done fairly well.

We are also now seeing what are called "flash sales" (see Fig. 2.2) by companies such as Groupon that reach shoppers via e-mail and smartphones and offer sales at various retailers that are good for only a limited amount of time.

It seems likely that retailing will continue, at least for the near future, to be led by large mass merchandisers that have highly efficient supply chain and logistics processes; specialty retailers that have great selection, customer service, and shopping experiences; and e-commerce [Smyyth LLC, 2011].

Figure 2.2 Flash sale example.

Retail's Value in the Distribution Channel

As shown in Fig. 2.1, retail is the last stage in the supply chain before products or services get to the customer. As a result, retail provides a variety of functions that add value for the customer. As we'll be discussing Lean in this book, it is important to understand what adds value for the customer so that we can also focus on the identification and elimination of non-value-added activities.

Value as a Utility

As both retailers and wholesalers are intermediaries and provide a utility or value to the customer, it is useful to look first at the value that they provide from a theoretical utilitarian perspective. The utilities they provide include:

- *Form utility*—Performed by the manufacturers (as well as third-party logistics companies, or 3PLs, that perform value-added activities such as kitting and display assembly) as they make the products useful.
- *Time utility*—Having products available when needed.
- *Place utility*—Having items available where people want them.
- *Possession utility*—Making the transfer of ownership to the customer as easy as possible, including the extension of credit.
- *Information utility*—Opening two-way information flows between parties (e.g., the customer and the manufacturer).
- *Service utility*—Providing fast, friendly service during and after the sale and teaching customers how best to use products. This is becoming one of the most important utilities that retailers offer.

Retail (and wholesale, upstream) provides all of these utilities to one degree or another.

Value as an Activity

In a more practical sense, retailers provide value in a variety of ways, including the following [Levy and Weitz, 2012]:

- *Assortment*—Individual manufacturers produce a limited number of items, whereas supermarkets can carry upward of 50,000 items. Thus, retailers are able to offer a wide assortment of products to consumers at one location.
- *Sorting or breaking bulk*—Manufacturers can achieve economies of scale in both production and transportation costs by shipping in large quantities. When applicable, wholesalers can offer products to retailers in smaller quantities, and in all cases, retailers can offer them to consumers in smaller quantities.
- *Hold inventory*—Retailers hold inventory so that it is readily available to consumers who have limited space in their homes.
- *Provide services*—Retailers provide value-added services such as credit, displays to test, product demonstrations, and salespeople to answer questions.

Wholesalers also offer value in a similar fashion, albeit upstream, to their retailer customers.

All of these activities provide value in the eyes of the customer and will help us to focus on what adds value and what does not later in the book.

Vertical Integration

In the supply chain, manufacturing, wholesale, and retail activities are typically carried out by different organizations. In some instances, however, there is vertical (backward or forward) integration.

In retail, an example of backward integration occurs when a retailer operates its own distribution centers to supply its stores. (This is fairly common with medium- to large-sized retailers.) Backward integration also can occur when a retailer has some manufacturing (owned or contracted) or wholesaling activities for private-label items.

A forward integration example would be when a manufacturer operates its own retail stores, such as Nike outlet stores, or a retailer operates a separate return center.

Value Chain

Vertical integration can be an ideal way for a retailer to increase the effectiveness of its "value chain" (the name of a model originated by Michael Porter that shows the value-creating activities of an organization; see Fig. 2.3) [Porter, 1998].

Activities That Add Value

In a value chain, each of a firm's internal activities listed here adds incremental value to the final product or service by transforming inputs to outputs (also see Fig. 2.3).

- *Inbound logistics*—Receiving, warehousing, and inventory control of input materials.
- *Operations*—Transforming inputs into the final product or service to create value.
- *Outbound logistics*—Actions that get the final product to the customer, including warehousing and order fulfillment. (Although this model is more geared toward manufacturing, this activity would also include delivery of retail products, similar to what we described in the SCOR model in Chap. 1.)
- *Marketing and sales*—Activities related to buyers purchasing the product, including advertising, pricing, distribution channel selection, and the like.

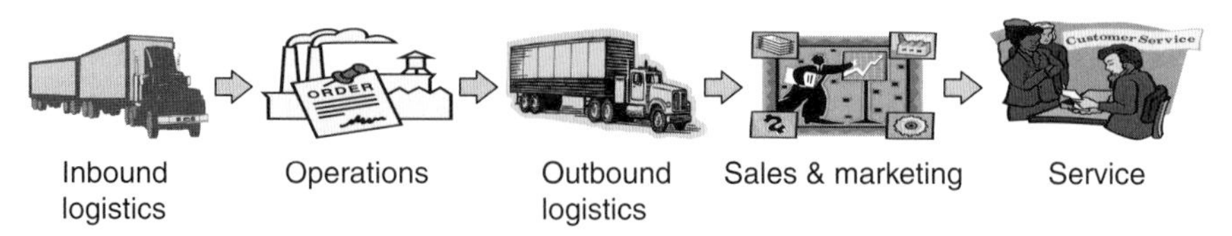

Figure 2.3 Primary value chain activities.

- *Service*—Activities that maintain and improve a product's value, including customer support, repair, warranty service, and the like.

There are also support activities identified by Porter that can add value to an organization, which are:

- *Procurement*—Purchasing raw materials and other inputs that are used in value-creating activities.
- *Technology development*—Research and development, process automation, and similar activities that support value chain activities.
- *Human resource management*—Recruiting, training, development, and compensation of employees.
- *Firm infrastructure*—Finance, legal, quality control, and the like.

As we can see, vertical integration as illustrated by the value chain model can be used to increase value and decrease waste in a firm's *entire* value system, including upstream suppliers and downstream buyers. This emphasizes the importance of collaboration among supply chain members, which will be discussed throughout this book.

Strategies for Tough Times

As it appears that we are in for many years of slow growth and growing competition, many retailers are struggling to come up with strategies that will help them succeed, many of which are good ideas but a piecemeal effort, to say the least. The various strategies as laid out by Berman and Evans in *Retail Management: A Strategic Approach* [Berman and Evans, 2012] include:

- *Rethinking existing store formats*—In urban areas, Walmart has used a strategy of smaller-format stores. To succeed with this strategy, Walmart reduces product assortments and maximizes its supply chain efficiencies, so that stockouts are minimized despite lower in-store inventory levels. It also has a "site-to-store" program that allows customers to order goods online (where Walmart has a much wider product offering), and then have the selected goods delivered to a local store for customer pickup.
- *Increased use of "pop-up" (or temporary) stores*—This is a relatively new phenomenon, but it has increased in popularity because of high retail vacancy rates and bargain-hunting shoppers. It's not just for holiday shopping, as pop-up stores may also carry manufacturers' overstocks, discontinued merchandise, and designer samples.
- *Low inventory levels to reduce markdowns*—Especially in a bad economy, retailers need to avoid high markdowns. They can do this by looking for special deals, closeouts, and the like. Some retailers such as Foot Locker and Champs are expanding their inventory at a slower rate and focusing their purchases on more popular items.

- *Increased promotions using coupons*—Coupons distribution and consumer usage are on the rise.
- *Shopper discounts based on credit card purchases*—Store-branded credit cards are offering discounts or rebates that also tend to increase purchases and visits.
- *Beginning the holiday season earlier*—We've all noticed that Christmas, Thanksgiving, and even Halloween shopping seems to start earlier each year. As retailers order products six or more months ahead anyway, when this is combined with a sluggish economy, holiday sales are starting earlier each year, it seems.
- *Reintroducing layaway plans*—Layaway plans have been around for many years, but they are making a comeback to both stimulate sales and offer credit to cash-starved shoppers. In this type of plan, a customer pays the product's total cost (usually with a small fee) in installments before being allowed to take the item home.

While these are great strategies for some retailers, in this book, we will outline a holistic approach to the identification and elimination of non-value-added activities that can be implemented as an integrated, long-term process to ensure that you're not just "rearranging the deck chairs on the *Titanic*," as they say.

This is not to say that there haven't been any "bright lights" when it comes to implementing some Lean concepts (albeit piecemeal) in retailing. According to Abernathy et al. in their book *A Stitch in Time: Lean Retailing and the Transformation of Manufacturing—Lessons from the Apparel and Textile Industries* [Abernathy et al., 1999], there have in fact been some Lean innovators in mass merchandising and department stores.

In mass merchandising, they point out how Walmart has successfully reduced its costs by using new technologies to track customer sales and inventory throughout its supply chain, thus enabling consumer demand to "pull" its orders.

In the case of department stores, they use Dillard's and Federated as examples of Lean retailing. They pioneered technology similar to what Walmart did in its stores, distribution centers, and headquarters to communicate internally as well as with suppliers. The major difference was that Dillard's and Federated carried almost 1 million stock keeping units (SKUs) at a store, whereas a mass merchandiser might carry around 125,000 items.

However, as Abernathy et al. point out, although Walmart has received the lion's share of the attention in this area, a variety of retailers have adopted various innovations (many going under the heading of "quick response," which will be discussed later) that have helped to get suppliers more involved in fast replenishment and reduced lead times starting in the early 1990s.

Now that we've looked at how retailing has evolved over the years, as well as where it adds value to the customer, we will examine the evolution of Lean from goods to service to see how it got its start in retail and wholesale and where it's headed.

CHAPTER 3

The Lean Journey: From Goods to Services

The term *Lean* didn't actually come into being until the publication of *The Machine That Changed the World* by James Womack, Daniel Jones, Daniel Roos, and Donna Carpenter in 1990. This book popularized the term in analyzing the future of automobile production by trying to understand how the Japanese automobile industry had simultaneously been able to improve quality and lower its costs much more quickly than Western companies [Womack et al., 1990].

The fact is, however, that many of the concepts and tools that fall under the "Lean" umbrella had already been around for quite a while.

Evolution of Lean

In order to understand Lean and its current and future applications, it is important to first briefly review the history of manufacturing (Fig. 3.1) and how the concept of Lean originated and evolved.

As mentioned previously, Lean is a team-based form of continuous improvement that is focused on the identification and elimination of waste from the customer's perspective. Obviously, things haven't always been done that way. If we look back to the start of manufacturing hundreds of years ago, most goods were made by individual craftspeople or artisans.

Early concepts like labor specialization (discussed by Adam Smith), in which an individual was responsible for a single, repeatable activity, and standardized parts (associated with Eli Whitney) helped to improve efficiency and quality. Up until that point, the individual craftsperson had made most, if not all, of an individual product (furniture, wagons, or something else). If a wagon wheel broke, it had to be made from scratch and might not even be exactly the same as the wheel it replaced.

Around the turn of the twentieth century, the era of scientific management began, in which concepts such as time-and-motion studies (emphasized by Frederick Taylor) and Gantt charts (developed by Henry Gantt) allowed management to measure, analyze, and manage activities much more precisely.

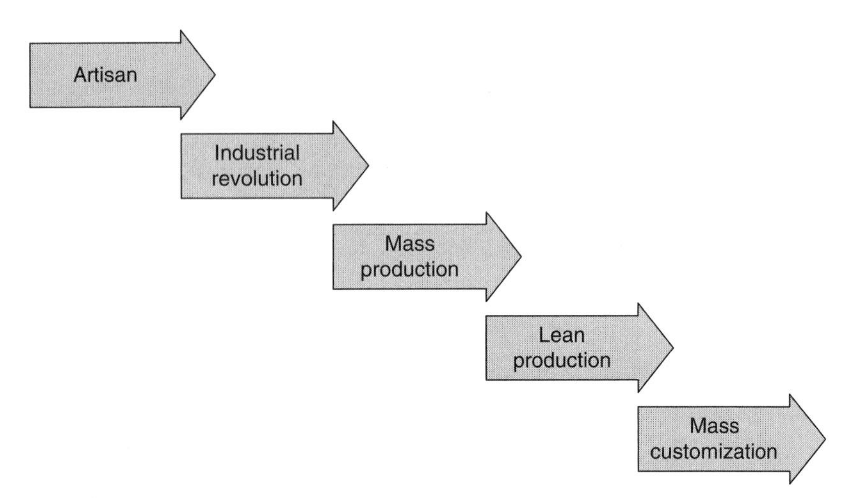

Figure 3.1 History of manufacturing.

The really big advance came in the early 1900s with the era of mass production. Concepts like the moving assembly line (e.g., Ford Motor Company), economies of scale (producing large quantities of the same item to spread fixed costs), and statistical sampling were applied. Today, this concept is known as a *push* process, which is the antithesis of today's demand *pull* (by the customer) Lean thinking.

As the saying used to go, you could have any color Ford Model T you wanted, as long as it was black! In a push process, goods are produced in advance of actual demand and kept in inventory (typically based upon some kind of forecast) in the hope that customers will buy them.

In the 1980s, we started hearing about a new concept called just-in-time (JIT). This concept actually originated out of necessity after World War II, when Japanese resources were scarce. It is a method of keeping a minimal inventory of material (or information)—not too much and not too little. The demand for JIT inventory is determined by the downstream activities (ultimately the customers) that use it. A great tool to implement this concept is *kanban*, which will be discussed later in the book, but in general terms, kanban is a visual method of replenishing inventory that is withdrawn or consumed by a downstream process.

JIT, kanbans, and other concepts and tools, such as total quality management (TQM), electronic data interchange (EDI), and employee empowerment, emerged. Many of them came from Japan, using statistical and other concepts taught to the Japanese after World War II by Americans such as W. Edwards Deming and Joseph M. Juran. The results were evident, as Japanese products gradually went from being poorly made (e.g., cheap toys and transistor radios in the 1960s) to high performance and quality (e.g., Toyota, Nissan, and other such companies had a reputation for quality from the 1970s onward).

The Japanese also laid the groundwork for the concept of "demand pull" or Lean systems (sometimes referred to as flow manufacturing, flexible manufacturing, JIT, or other terms) to emerge. The true precursor to this is the much-written-about Toyota Production System (TPS), which was started in the late 1940s in Japan and which focuses on continuous improvement and respect for people. The objectives are to design out overburden (*muri*), inconsistency (*mura*), and waste (*muda*). TPS encourages employees to get to the source of a problem or issue by focusing on waste (which will be discussed in more detail in the next chapter) [Myerson, 2012].

Manufacturers and Wholesalers Partner with Retailers

In the early 1990s, the term *Quick Response* (QR) was created to describe an organizational focus on shortening lead times and having a "quick response" to demand for existing and new products as well as design changes.

Specifically, QR was initiated by the U.S. textile Industry in 1984 as a method for improving efficiencies in manufacturing and supply chain processes that crossed organizational boundaries.

It was driven by the fact that traditional fashion markets typically have two seasons per year, but there has also been an increasing need for a faster change of ranges, styles, and colors. So QR gives the fashion industry an ability to scale up (or down) quickly, along with being better able to consider and include consumer preferences in the design process itself.

QR is a just-in-time inventory partnership strategy between suppliers and retailers of general merchandise involving supply techniques that allow retailers to adjust their demand in response to fashion trends and seasonal sales variation. The main objective of QR is to respond quickly to market changes and cut ordering lead times. It is focused on agility, which can lead to efficiency throughout the supply chain.

QR is enabled by technologies such as:

- *Electronic data interchange (EDI)*—A standardized format for businesses to use to exchange data electronically.
- *Bar codes*—Optical machine-readable data relating to the item to which the code is attached. Typically, a bar code contains the manufacturer and item code.
- *Radio-frequency identification (RFID)*—The wireless use of radio-frequency (RF) signals to transfer a multitude of data to automatically identify and track tags that are attached to objects.

These and other enabling technologies will be discussed in more detail later in this book.

In many cases, QR is accomplished through automatic replenishment (AR) or, at a minimum, with the help of the suppliers (i.e., manufacturers or wholesalers).

By using a tracking system for inventory, the retailer's supply of products is replenished automatically when stocks run low. AR systems will also alert stores to problems. The automatic ordering and replenishment is typically based on previous retail sales or retail warehouse withdrawals (or a combination of both). Automatic replenishment relies on the supply source being able to respond directly to demand.

Sales information is transmitted directly to the supplier, who replenishes products directly to the store (or the retail distribution center). Usually, the supplier replenishes according to a predetermined assortment plan for the season. QR is supported by flexible manufacturing techniques that can switch to making fast-selling merchandise instead of slower-moving lines.

QR has general benefits, such as:

- Quicker deliveries, with shorter lead times from order receipt to delivery
- Faster inventory turns and lower inventory investment
- Increased cash flow
- Improved communications through the establishment of a strong partnership

The retailer, in particular, receives many benefits, which include:

- Improved communications
- Improved planning systems
- Quicker access to sales information
- Easy tracking of products
- Reduction of stock holdings
- A higher level of sales because of fewer stockouts
- An improved profit margin
- Enhanced customer satisfaction

Around the same time that QR came into being, the food distribution industry came up with Efficient Consumer Response (ECR).

ECR is a strategy to increase the level of service to consumers through close cooperation among retailers, wholesalers, and manufacturers. It aims to improve the efficiency of a supply chain as a whole, beyond the wall of retailers, wholesalers, and manufacturers, to gain larger profits than each of them pursuing its own disparate business goals could achieve.

Specifically, ECR attempts to place orders more frequently in smaller lots to shorten lead times and reduce inventory costs. In my experience, in many cases, as with QR, the supplier takes responsibility for monitoring and replenishing the retailer's distribution center inventories with the approval of the retailer.

Another more recent form of collaboration between suppliers and retailers is CPFR® (Collaborative Planning, Forecasting and Replenishment)*, which was originated jointly by Walmart and Warner Lambert in the 1990s. In CPFR, both

*CPFR® is a registered trademark of the Voluntary Interindustry Commerce Standards (VICS) Association.

trading partners develop a joint business plan, including a promotion calendar, and conclude with a joint replenishment plan. The retailer and manufacturer also agree on a joint sales and order forecast.

CPFR differs from the other collaborative programs in that both parties are informed of exceptions, generating collaborative activities that help to resolve these exceptions. There is also a reliance on using the exceptions to point out discrepancies, when used with a large number of stores and many items. CPFR also utilizes Internet-based technologies to collaborate on planning and execution, allowing trading partners to agree on a *single* item forecast to drive production and distribution, rather than the typical method where each partner has its own *different* item forecast.

QR, ECR, and CPFR help today's retailers to more efficiently manage the exploding number of items that they must stock in order to be successful. Forward-looking manufacturers also jumped on board these programs quickly, as they took a long-term view of the benefits of partnering with customers, which allows them to gain visibility and efficiencies in their own production and distribution processes as well as the entire supply chain both downstream and upstream.

Lean Office

Around the year 2000, Lean techniques began to move from the shop floor to the office, as it became apparent that waste was everywhere and that offices had some of the same characteristics as manufacturing, such as batching, setups, equipment failure, standardized work, and the like. In fact, as much as 60 to 80 percent of the lead time for a product or service can be found in the office environment [which may include functions as diverse as customer service, order management, quoting, engineering, and research and development (R&D), to name a few].

The benefits of a Lean office vary, but include:

- More flexibility and responsiveness
- Reduced lead time
- Reduced errors
- Reduced extra processing
- Improved utilization of personnel
- Reduced transactions
- Simplified processes [Myerson, 2012]

Lean Supply Chain and Logistics Management

It has only been in the past 5 years or so that the concept of Lean has moved to the supply chain and logistics management environment. I believe there are a number of reasons for this.

A major reason is that, as previously mentioned, Lean started in manufacturing (especially repetitive, assembly-line manufacturing), then gradually moved to other manufacturing processes, such as continuous flow (e.g., chemicals, foods, and beverages) and, to some degree, batch processing or job shop production (smaller, often customer-specific production). Most manufacturers (rightly or wrongly) wanted to "Lean out" within their own four walls before working heavily with customers and suppliers. So, in a way, moving to the supply chain and logistics area is a natural evolution. Companies now realize that they can take things only so far without collaborating and partnering more closely with their customers and suppliers. Otherwise, in many cases, they just push their inefficiencies on to suppliers (e.g., JIT of raw materials) and are constantly frustrated by distorted and volatile customer demand (known as the "bullwhip effect").

Additionally, while there has always been an emphasis on reducing costs, the recent economic meltdown has focused the red-hot light on supply chain management even more. So these days, any tool that can wring inefficiencies out of a system draws people to it [Myerson, 2012].

Integrated and Agile Supply Chain

Ultimately, as suggested by A. Harrison, M. Christopher, and R. I. Van Hoek (1999; Fig. 3.2), an agile and Lean supply chain must be:

Market sensitive—Closely connected to end-user trends. Point of sale (POS) data are analyzed daily and then used to determine replenishment requirements.

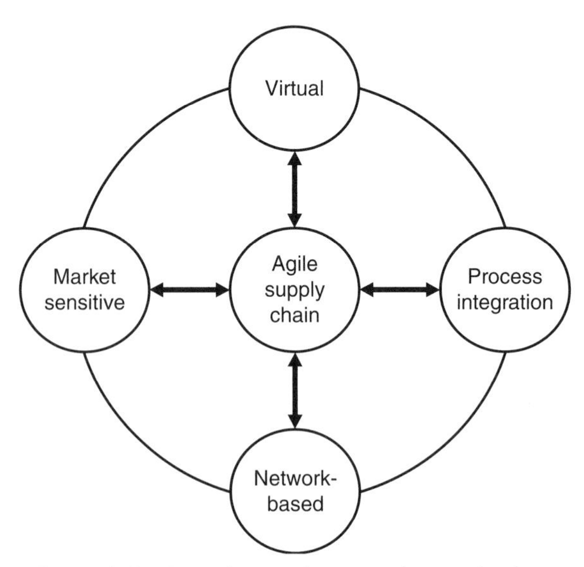

Figure 3.2 Foundations for an agile supply chain.

Virtual (and collaborative)—Relies on shared information across all supply chain partners. The supply chain is virtual in that it is connected and integrated through shared information on actual demand so that all the members of the supply chain, from manufacturers to wholesalers and to retailers, are seeing the same set of figures. In the past, few retailers would share POS data with their suppliers. Shared information can result in higher levels of retail availability with lower inventory levels throughout the supply chain. Transaction costs can also be reduced if comanaged inventory methods such as Quick Response are used.

Network-based—Has flexibility by using the strengths of specialist members and is flexible enough to cope with sudden changes in demand.

Process integration—Includes a high degree of process interconnectivity among the network members, both downstream (toward customers) and upstream (toward suppliers and vendors). This includes the capability to create almost seamless connections with supply chain partners, with no delays caused by handoffs between the different stages in the chain, and with transactions tending to be paperless [Harrison et al., 1999].

Current State of Lean

While many retailers, starting in the 1990s, collaborated and shared information with suppliers, helping to motivate them (and themselves) to become more efficient, especially in terms of inventory, this was only the beginning of implementing a Lean philosophy, especially for small- to medium-sized retailers (and wholesalers).

In the next chapter, we will explore some Lean success stories in retail and wholesale to get a better feel for where we are today and where Lean might fit in the future (at least strategically) before discussing Lean concepts, tools, and methodologies to truly transform an entire retail organization well for the long term.

CHAPTER 4

Lean Retail and Wholesale: Early Signs of Promise

As we mentioned, retailers were in some ways the drivers of Lean manufacturing in the 1990s with the advent of Quick Response (QR), Efficient Consumer Response (ECR), and Collaborative Planning, Forecasting and Replenishment (CPFR) programs. In fact, probably more than most industries, retail has used technology to help improve the bottom line.

In the 1980s, retail adopted the Universal Product Code (UPC) and the use of bar codes with store point of sale (POS) systems to increase efficiency and accuracy.

The 1990s saw some major retailers implement some planning and optimization tools for the forecasting, merchandising, warehousing and distribution, and pricing functions.

Now, in the 2000s, there is more supply chain integration, sophisticated retail enterprise resource planning (ERP) systems, e-tailers, more pricing and markdown optimization tools, and many specialized applications such as retail product lifecycle management systems (known as PLM) that support design and development processes for better integration with merchandising, assortment and line planning, and sourcing.

So before we go on to discussing where retail and wholesale can further benefit from a true "end-to-end" Lean philosophy, let's examine some "best-in-class" examples of Lean in retail and wholesale.

We will organize these examples using the SCOR model components discussed in the first chapter: Plan (retail strategy), Source or Make (merchandise management), Deliver (warehouse and store operations), and Return (reverse logistics).

Plan

There is perhaps no better example of Lean planning in retail than Walmart's Retail Link program. Started in 1992, Retail Link is a system that is designed to strengthen supplier partnerships. The system provides vendors with information on sales trends and inventory levels.

> Walmart implemented the first companywide use of UPC bar codes, in which store level information was immediately collected and analyzed, and the company devised Retail Link, a mammoth Bentonville database. Through a global satellite system, Retail Link is connected to analysts who forecast supplier demands to the supplier network, which displays real-time sales data from cash registers and to Walmart's distribution centers (DCs). [This includes] information from point-of-sale data, the cash register that they put into their system and share with all their partners.

What made this so innovative at the time was that

> A lot of companies weren't sharing that [POS data]. In fact, they were using third parties where they had to pay for that information.
>
> Walmart's approach meant frequent, informal cooperation among stores, distribution centers and suppliers, and less centralized control. Furthermore, the company's supply chain, by tracking customer purchases and demand, allows consumers to effectively pull merchandise to stores rather than having the company push goods onto shelves [Traub, 2012].

I had firsthand experience with this in the 1990s when I was with Church & Dwight (the maker of Arm & Hammer), and I can attest to the fact that it was a true collaborative approach that benefited both the customer (Walmart) and the supplier (us). Besides the company's sharing of POS data with us, we would meet in Bentonville, Arkansas, on a regular basis to discuss forecasts for our various products that Walmart carried.

Because forecasting is not an exact science, some manufacturers like Hewlett-Packard (HP), working with distributors of their products, have used the concept of *postponement* to increase the efficiency of the supply chain by moving product differentiation closer to the end user. In HP's case, the company redesigned its printer supply chain to overcome the problem of variability in demand by moving the product differentiation point to the distribution centers.

In redesigning the printer, HP first produced the standard common core printers without a power supply, with final power supply assembly being delayed and finished by the distributors.

Source or Make

Efficient Consumer Response (ECR) programs were implemented at numerous grocery chains throughout the 1990s, including H. E. Butt, Shop-Rite, and Spartan Stores (I was involved with two of these chains while I was at Church & Dwight).

Quick Response (QR) programs were successfully implemented at numerous apparel and mass merchandisers such as Dillard's, Kmart, and J. C. Penney (I worked with Kmart on a QR project while I was at Church & Dwight).

At one point, in fact, the consulting firm Kurt Salmon and Associates (1993) had projected a $30 billion savings from ECR strategies.

Walmart has utilized vendor-managed inventory, or VMI (also known as QR and ECR, especially in grocery and retail), extensively, where manufacturers become responsible for managing their products in Walmart's warehouses. As a result, Walmart has close to 100 percent order fulfillment on merchandise when using VMI programs with vendors [Traub, 2012].

Various pilot CPFR projects that included retailers such as Wegmans, Kmart, Walmart, Target, Tesco, Meijer, and Sainsbury's achieved 30 to 40 percent improvements in forecast accuracy, significant increases in customer service, sales increases of between 15 and 60 percent, and reductions in days of supply of 15 to 20 percent.

Somewhat confirming the results of these projects, Advanced Market Research (AMR) Research reported in 2001 on the range of results actually achieved by many early adopters of CPFR. Retailer benefits included:

- Better store shelf stock rates: 2 to 8 percent
- Lower inventory levels: 10 to 40 percent
- Higher sales: 5 to 20 percent
- Lower logistics costs: 3 to 4 percent [Sheffi, 2002]

Why VMI in Retail?

Fluctuations in demand are a major issue for the supply chain. They can affect service and, ultimately, profits. In retail, management actions such as orders driven by general unpredictability, promotion plans, poor communications, and shortages can all contribute to erratic orders.

The supply chain is a major cost center for manufacturers, ranging from 50 to 80 percent of the cost of sales (this varies by industry), and a major cost center for retailers as well, as they typically pay for inbound transportation as well as the inherent carrying costs of acquiring and holding inventory (this ranges from 15–40 percent of the value of inventory on an annual basis).

As a result, it is typically easier to reduce costs by a relatively small percentage and get a contribution to profit equivalent to that from increasing sales by a much larger percentage. For example, a company with a 10 percent profit margin and a supply chain cost of 60 percent of sales would need to increase sales by $4 to have the same impact on the profit margin as a $1 supply chain cost reduction, and we all know how hard it is to increase sales in the current economic environment.

Another reason is operational. There is something called the "bullwhip effect" (see Fig. 4.1). Basically, it describes the magnified effect (especially on inventory, operational costs, and customer service) that occurs when orders move up the supply chain. This can be caused by a variety of things, such as forecasting errors, large lot sizes, long setups, panic ordering, variance in lead times, and the like [Myerson, 2012].

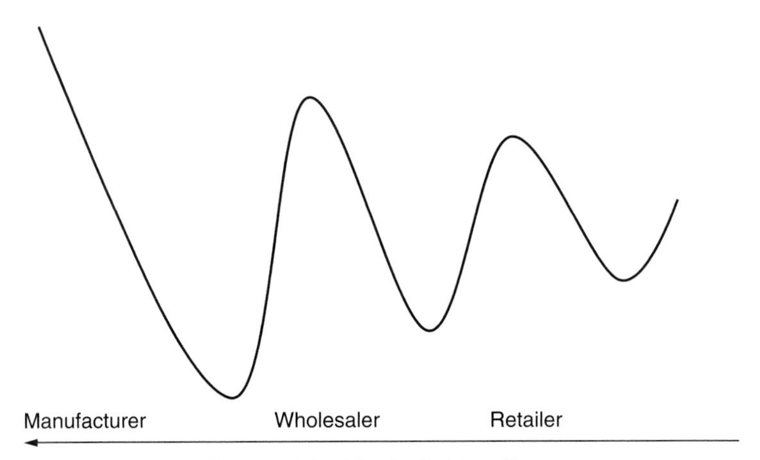

Figure 4.1 The bullwhip effect.

VMI, also frequently referred to in retail as QR and ECR (all three of which were described earlier), can help to reduce demand fluctuations and thereby reduce cost. These more predictable order patterns, which result from better communication and collaboration among manufacturers, wholesalers, and retailers, can smooth demand and reduce the bullwhip effect to some extent, thereby reducing the need for buffer or safety stock inventory requirements to some degree.

Infrequent large orders from retailers require suppliers to keep excessive finished goods inventory. VMI reduces the peaks and valleys of production, somewhat reducing the need for excess capacity and inventory.

Buyers in retail frequently have conflicting performance measures. Typically, they are measured at month-end with regard to inventory levels and customer service, which in some cases can be conflicting. Buyers must allow inventories to drop at month- or quarter-end to meet their targets, and this may both negatively affect customer service and cause a large spike when reordering. VMI results in smaller orders being placed more often, which tends to smooth demand upstream for suppliers. As lower safety stocks are needed, VMI tends to reduce inventory costs for retailers and suppliers while simultaneously increasing service levels by having the correct product available when needed. Production becomes more predictable, reducing manufacturing costs through better use of capacity and lower inventory levels.

VMI allows for greater coordination among different buyers, which helps the suppliers' need for smoother, more predictable production.

Transportation costs may spike at first because of the smaller order sizes, but if it is executed properly, VMI may actually reduce transportation costs. The supplier can better coordinate and consolidate the supply process among its many orders and items.

In retail, service is usually measured in terms of product availability. Lost sales, therefore, are to be avoided at all costs, especially during a promotion.

Thus, supplier dependability is critically important, and those suppliers that are viewed as being dependable tend to be favored with better shelf space given to them.

The supplier, of course, benefits from greater revenues by minimizing lost sales and improved store presentation.

A VMI program will help improve the coordination of replenishment orders and deliveries to improve service. VMI reduces the need for "firefighting" through better dependability, which also reduces inflated orders as a result of this panic [Waller et al., 1999].

Deliver

Cross-docking is the practice in logistics management of unloading materials from an incoming tractor trailer and loading them directly into outbound trucks, with little or no storage in between the two. Typically, the longest that material will sit in a cross-dock facility is 24 to 48 hours.

Walmart uses this strategy to gain a huge competitive advantage over its competitors by streamlining its supply chain. For Walmart and some other major retailers, cross-docking has reduced handling, operating, and inventory storage costs. It has also helped them to deliver products to the final customer faster.

At the store and distribution center (DC) level, the primary emphasis of Lean has been focused on inventory, as previously noted. There has also been some use of the Lean concept of 5S, or workplace organization, which we will describe later in the book.

As it's easy to understand and implement, 5S is typically one of the first Lean tools used. Retail stores, in general, are fairly neat and organized, but 5S is all about taking this to the next level.

I've had several university students in my classes who have worked at some major retailers. In one case, the student worked at Burlington Coat Factory (BCF). She presented a discussion of a successful 5S pilot project at a store. She mentioned the drastic difference between "before" and "after" in terms of neatness and organization in both the front and the back room of the store. However, from what she said, the pilot was only temporary and was never rolled out. Such is the fate of Lean initiatives in many cases where the Lean philosophy doesn't take root.

Another student worked at Wegmans, which is one of the few grocery retailers that I know of that has an ongoing continuous improvement program. The student presented a project detailing the 5S activities at the firm's stores. It was very impressive and included many of the tools necessary for it to sustain the program for the long term, such as audits, checklists, and so on.

A 2009 *Wall Street Journal* article documented that 24 percent of the annual revenue for Starbucks is store labor, and that, as a result, there was room for improvement to lower that number, as 30 percent of the store employees' time is motion-related—walking, reaching, and bending. Starbucks used a variety of basic techniques, resulting in the reduction of the time required to make coffee from more than a minute to as little as 16 seconds. These techniques included:

1. Moving items closer to where they were used reduced excess movements behind the counter.
2. The order of assembly of the coffee was revised (there are 80,000 combinations of drinks available!).
3. Commonly used syrups were located nearby in a reachable location.
4. Whipped cream, chocolate, caramel, and the like were moved closer to the delivery area, as adding them was the last step before serving up the coffee [Jargon, 2009].

I have delivered some Lean training at several large and highly automated retail distribution centers in New Jersey. We were able to create and implement 5S in areas such as receiving, picking, and ticketing by creating a variety of visual job aids that were easily understandable and therefore easy to follow. They included digital photographs and instructions in English and, in some cases, Spanish (Fig. 4.2).

It seems apparent that the concept of Lean at the retail level, at least in the United States, is in its infancy. From a review of articles, it appears that it is a bit further along in Europe and Australia (although not that much further along, as basic concepts such as 5S are primarily used).

Once such example is Sonae Modelo Continente (MC), a global leader in the retail sector, which implemented 5S, giving workers a structure for better organizing their workplace, including stocking and receiving areas. Another concept that the firm implemented was the visual workplace—the implementation of labeling systems, markings, color codes, and other visual tools to provide visual guidelines for an orderly environment. Finally, it used standardization where possible, which is the creation of written (and visible) processes that allow the organization to sustain the gains made through kaizen events (i.e., process improvement projects). All of this information was shared with other divisions.

Visual management, 5S, and standardization allowed Sonae to increase productivity so that it could offer its clients better service [Kaizen Institute Ltd., ; accessed 2013].

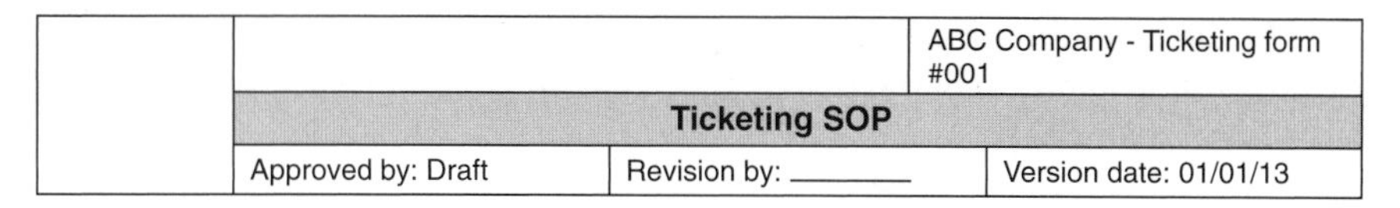

			ABC Company - Ticketing form #001
	Ticketing SOP		
	Approved by: Draft	Revision by: ______	Version date: 01/01/13

Phase	Step	Comments	Visuals
1. Start up	• Production sheet	• Fill out with name and work center code	
	• Login gun	• Get RF Gun • Login RF Gun • Scan RF Gun to ticket machine for specific lane	
2. Select and scan-in carton	• Select "CTN" (#4: Carton Number) on gun		
	• Scan carton bar code	• Ticket prints on machine	
3. Print and prepare tickets	• Get tickets • Match "CTN"	• CTN on tickets should match CTN on carton label	
	• Check case number on tickets vs. case number on carton and style and color	• Check ABC Label label or Vendor label on carton • Match style and color on ticket vs. label	
4. Seeding	• Cut open box in middle • Insert tickets • Seal carton with "Tickets Enclosed"		
	• Sign off • Push box on "outbound" ticketing line	• Write employee number on "Disposition Label" on outside of box	
5. Update production sheet	• Enter number of cartons seeded		

Figure 4.2 Visual job aid.

E-commerce

We can't neglect the emergence of e-commerce as a form of Lean delivery system, as it can replace (or at least supplement) "brick-and-mortar" retail operations to improve efficiencies and lower costs.

While e-commerce is big ($176 billion a year in sales in 2010), it is still a relatively small portion of retail sales (8 percent in 2010), but it is growing at a fast clip (around 9 percent annual growth rate) [Forester, 2011]. A lot of this growth, of course, is coming at the expense of shopping malls, department stores, and electronics stores.

E-commerce companies typically have lower operating expenses than offline retailers, and therefore in many cases have lower prices than brick-and-mortar retailers. However, this isn't always the case, as in many cases, especially in start-ups, e-commerce businesses buy in smaller quantities, resulting in a higher cost of goods sold, and also often offer free shipping and free return shipping, which can drive up fulfillment costs.

One key trend to note, and keep an eye on, is that e-commerce appears to have gravitated toward wealthier customers and higher-margin products in recent times, at least partially as a result of the most active online buyers tending to be the most affluent [J.P. Morgan, 2010].

Return

Reverse logistics is the process of moving goods from their final destination in order to capture some value or, at least, for proper disposal. Remanufacturing, refurbishing, and repair activities can also be included in this definition.

More specifically, reverse logistics can include the processing of returned merchandise because of damage, excess seasonal and regular inventory, salvage, recalls, recycling, hazardous material programs, obsolete equipment disposition, and asset recovery.

As margins are razor slim, good return management is extremely important to retailers, as returns reduce the profitability of retailers by 4.3 percent on average. Additionally, some firms in the retail industry have used it as a competitive weapon through liberal return policies (e.g., Zappos.com offers free shipping on returns).

As a result of its potentially high impact, retailers are now paying more attention to reverse logistics, especially in the area of technology. In fact, retailers have made large investments in technology to improve their reverse logistics systems, and most manufacturers lag behind retailers in almost every technology category.

Grocery retailers led with early innovations such as reclamation centers, leading to the establishment of centralized return centers, which in turn have resulted in benefits for most firms that have implemented them.

According to a study by the Reno Center for Logistics Management at the University of Nevada, nearly twice as many retailers as manufacturers included in the research used automated material-handling equipment. The survey found that retailers are more likely than manufacturers to use bar codes, computerized return tracking, computerized returns entry, electronic data interchange (EDI), and radio-frequency (RF) technology to enhance their reverse logistics management process [Hawks, 2006].

Perhaps because of the nature of the Internet, e-tailers like Zappos have been leading the way in this area:

> Zappos customers can order any styles, colors, and sizes, try them on, and then return the ones they don't want for free. In fact, the company encourages it. All shoppers have to do is use UPS Returns® on the Web and call a driver for pickup. UPS handles the rest.
>
> This seamless shipping experience is one reason 75 percent of Zappos shoppers are return shoppers; after all, Zappos is all about an easy experience and instant gratification. That's how the company reached $2 billion in sales in 2008.
>
> And not only can customers easily return their purchases, Zappos can easily monitor the incoming returns and inbound goods from vendors, enabling them to plan staff accordingly and replenish stock for resale quickly. [UPS, accessed 2013]

Even in a Lean supply chain, returns are unavoidable. Returns reduce the profitability of retailers substantially, so that good return management is critical.

Many retailers handle returns themselves, but some have found it useful to outsource the process, as it may not be a core competency and can require substantial investments in systems, processes, and technologies.

Electronic retailer Best Buy used software and services from third-party logistics (3PL) provider Genco Distribution System to integrate its centralized returns management process, resulting in a reduction of 50 percent in processing costs and a reduction in its internal returns management team from ten to two employees.

Returned products are brought to Best Buy stores, shipped to one of fifteen consolidated (and home delivery) centers, then sent to one of two Genco-managed return locations.

Road Runner Sports is a multichannel retailer of running shoes that has about 500 to 700 returns per day (about 140,000–150,000 per year out of 1 million packages sold). As the firm feels that customer service is a core competency, it implemented SmartLabel from Newgistics, which is a prepaid, preaddressed, barcoded return label that a customer can remove from the purchase invoice and put on the package, then drop off the package in any USPS mailbox. Returns are consolidated and passed on to Newgistics, with Road Runner then sending customers a return status e-mail. As a result, return status calls have dropped by 10 to 15 percent, and customers seem very satisfied with SmartLabel, as usage is 70 to 80 percent [Blanchard, 2005].

Where Do We Go from Here?

While the Lean activities discussed in this chapter have been helpful, they are for the most part concentrated in certain areas (such as supply) and used by only a relatively small segment of retail (i.e., large organizations).

In order to really make a significant impact, Lean requires a complete understanding of the basic concepts and tools along with a philosophy of continuous improvement that permeates throughout the organization using a holistic approach. That is what we will discuss next.

PART II

Future State

CHAPTER 5

Thinking Lean: Basic Lean Concepts and Tools—Building a Solid Foundation

What Is "Waste"?

Many people think they already know what "waste" in a process is as a result of always hearing about government waste. While there are some similarities, it is not really the same. In fact, what we mean by waste is a different way of thinking to most people. The idea of thinking of a process in terms of value-added versus non-value-added or wasteful activities is easy to understand and definitely makes it easier to identify and eliminate waste.

The most common way to describe these non-value-added activities is by using the concept of the *seven wastes*, which we'll get to shortly. Some people like to add another waste (behavioral waste) when discussing the topic.

In general, waste can be defined as anything that does not add value to a process. Typically, when a product or information is being stored, inspected, or delayed; is waiting in line; or is defective, it is not adding value and is 100 percent waste.

The original seven wastes came from the Toyota Production System (TPS), discussed earlier. The seven wastes are unnecessary *transportation* or *movement*, *inventory*, excess *motion*, *waiting*, *overproduction*, *overprocessing*, and *defects or errors*. A good way to remember this is by the acronym TIM WOOD. You can also add the eighth waste of *behavior* (or underutilized employees) to this, as it can sometimes be the biggest waste of all.

These wastes are applicable to any process, whether it is manufacturing, office, retail, wholesale, or something else—essentially any process in your organization. What follows are descriptions of each of the wastes, as well as some examples.

Inventory Waste

It is best to go out of the TIM WOOD order when reviewing the eight wastes and start with inventory, as it is both the most visible and actually an end result of the other wastes.

Inventory is a buffer between suppliers, manufacturers, wholesalers, retailers, and customers and is needed to compensate for lead times (e.g., required for transportation, manufacturing, and so on) and variability in the system, such as forecasting errors, late deliveries, setup times, scrap or rework, quality problems, and downtime.

In an office environment, inventory might refer to information, such as customer orders, or supplies, but it is still just as important, as it directly affects the cycle time.

There are four kinds of inventory:

1. *Raw materials*—Typically purchased materials and components.
2. *Work-in-process (WIP)*—Products for which the transformation process has started but has not yet been completed. This can include finished goods that are awaiting final packing into kits or displays.
3. *Finished goods*—Finished, salable products.
4. *Materials, repairs, and operations (MRO)*—Inventory for equipment spare parts and supplies.

All of these types of inventory cost money to maintain. This is called *holding* or *carrying costs*. These costs can range from 15 to 40 percent of the value of a product and include the cost of capital (i.e., borrowing costs or opportunity cost lost if the money could have been invested elsewhere), taxes, storage, insurance, handling, labor, obsolescence, damage, and pilferage.

For example, if a business bought $100,000 of a raw material, had holding costs of 30 percent, and did not use the material for a year, the material would have actually cost the business $130,000.

In reality, businesses need some inventory and typically have to balance the trade-off between the cost of carrying inventory and customer service when determining how much.

There is the old analogy of a boat on the water, with the level of the water representing the amount of inventory and the jagged rocks below representing variability (Fig. 5.1).

Excess inventory is really a symptom of the problem. It is often said that the idea is to lower the water level until the "rocks" show above the water. In fact, in many companies, the finance department will pass an edict to lower inventories by x percent by year-end, which may expose the rocks, but can also create significant customer service problems. It is perhaps more conservative, but more effective to take the opposite approach. Identify the sources of variability and then, using analytical tools (described in more detail in Chap. 6) such as the Pareto principle (also known as the 80/20 rule), root cause analysis, and the "five whys" (keep asking "why" until you get to the root cause), reduce or eliminate the variability, *then* reduce the inventory levels.

Figure 5.1 Need to reduce variability.

Transportation or Movement Waste

This type of waste can include transporting, temporarily locating, filing, stocking, stacking, or moving materials, people, tools, or information. Ideally, when material is received, it should be touched only twice: once to put it away and another time to pull it for consumption. However, the reality is that things rarely happen this way.

Material may be moved from one place to another on the floor, put on a storage rack, pulled to remove some material, then returned to a different rack, and the like. All of this excess movement is wasteful. Not only are companies paying a forklift driver to move the material, but each time it is moved, damage may occur, and each time it is moved, inventory accuracy may be affected. When material is returned to a different spot, there is the risk of losing it and accidentally ordering more (yes, that does happen).

In many cases, transportation waste can be something very obvious that you have just learned to live with, like the copier being too far from your desk, paper and staplers being kept too far away from the copier, no signs identifying areas or departments, or just simply poor office layout.

That is why the foundation concepts of layout and the visual workplace are so important, and these will be described a little later in this chapter.

When looking at a layout, think of *flow*. This is important whether we are talking about a manufacturing facility, an office, a retail operation, or a warehouse.

Consumer transportation costs, which are a waste in the purest sense, have also been minimized by "super" grocery stores that include just about everything—banks, post offices, restaurants, video rental, and even nail salons all located within one store. This is a win-win situation, as consumers look at this as a convenience, and retailers look at it as an opportunity to keep you in the store longer.

Motion Waste

The concept of motion waste is best described by the idea of having the things you use more often closer to you (and at waist level) and the things you use less often further away and higher up. Any motion that does not add value to the product or service is wasteful.

The Lean concept of *point-of-use storage* is applicable here. It basically means having just enough material or information nearby, which can be replenished when needed from further away (a *kanban*, which will be discussed in Chap. 6, is an excellent visual tool for this type of replenishment).

Some examples of motion waste are looking for tools, excessive bending or reaching, and materials placed too far away.

When thinking about motion waste, the term *ergonomics* comes to mind. Ergonomics is the science of how humans interact with equipment and the workplace. So in terms of motion, you want to consider not just efficiency, but safety as well (i.e., avoiding back injuries, carpal tunnel syndrome, and similar problems).

Some grocery stores have attempted to minimize the waste of motion and transportation by placing milk, eggs, and other staples in easy-to-reach refrigerators in the front of the store. Most fruits and veggies are available cleaned and already cut up in packages, ready for cooking with almost no preparation. (This is quicker than purchasing the raw food item, but also more expensive.)

Waiting Waste

The waste of waiting is simply time spent waiting for materials, supplies, information, and people that are needed to finish a task. Everyone, whether on the shop floor, in a warehouse, or in an office, can easily identify with this type of waste. It is both frustrating and counterproductive.

In most processes, a great deal of a product's or service's lead time is spent on waiting. In many cases, the waiting is caused by the next operation.

This can be a result of long setup times, large batch sizes, and downtime. The result can be larger-than-needed amounts of WIP inventory. In an office environment, time can be spent waiting for equipment to start up, a printer or computer breakdown to be fixed, signatures, employees on different work schedules, and even meeting attendees who do not show up on time (which never happens, of course).

In many warehouses or distribution centers (DC), products can sit waiting between different steps in the process (e.g., receiving, putting away, replenishing, picking, packing, and shipping).

In the case of grocery and mass merchandise retailers, we find that the waste of waiting in a long line can be reduced with quick self-checkout lines, especially if you only have few items to purchase . . . with the added benefit of being able to bag your items the way you prefer.

Overproduction Waste

Overproduction, and its sinister sibling overprocurement, is manufacturing, ordering, or processing something before it is actually needed. This typically results in an excess of another major waste already mentioned, inventory (raw, WIP, finished goods, or MRO). In addition, it can result in longer-than-necessary lead times, higher storage costs, and potentially a greater number of defects (which may be harder to detect) because of larger-than-needed batch sizes. Overproduction inhibits the smooth flow of materials, a basic tenet of Lean. Instead of just-in-time (JIT), it ends up being just-in-case!

In the office environment, this may involve preparing or printing paperwork earlier in batches as a result of long setup times (yes, there are setups in the office as well as on the shop floor), preparing a report early or in its entirety instead of online or as an exception report, and memos and e-mails that copy "the world."

The warehouse may suffer from some of the office-type overproduction wastes, as well as from others like pulling orders earlier than needed or ordering supplies and packaging materials in large batch sizes.

Grocery stores have attempted to minimize the waste of overproduction by reducing the price of baked goods and produce that are about to expire. Grocers win because they still make a profit, just a smaller one. Consumers also win because they get a bargain on an item that they were probably thinking about buying anyway.

Overprocessing Waste

Overprocessing happens when too much time or effort is put into processing material or information that is not viewed as adding value to the customer. This can also include using equipment that may be more expensive, complicated, or precise than is actually needed to perform the operation.

This may occur when there are unclear customer specifications, when a product or service is continually refined beyond what the customer wants or needs, or when a lengthy approval process is involved.

Examples of this can be overpackaging (have you ever opened a Christmas toy for a child and wondered whether all of the packaging materials were really necessary, and also what a waste they are from an environmental perspective?) and overchecking.

In the office, overprocessing can include things like sending the same information in multiple formats (fax, e-mail, and overnight delivery), repeating the same information on different forms, reentering data, and unnecessary information on a form.

An example of overprocessing in retail would be when a store requires you to present your store membership card in order to get a discount on an item. In some cases, the store will allow you to just give your phone number and look up your card number when you don't have the card handy (as you may have many of them); in others, it may refuse to honor the discount without the physical card (I've experienced that frustration myself).

Defect or Error Waste

In manufacturing, the waste of defects primarily refers to repairing, reworking, or scrapping materials. The further along a product with a defect gets, the more costly it is to the company, as the company may need to rework it into the system, scrap it, and make it all over again, or, in the worst case, have it returned by the customer (which can include safety and liability issues, as in the Tylenol and Toyota recalls, for example). A lot of extra, non-value-added activities take place as a result, such as quarantining, reinspection, and rescheduling, possibly resulting in overtime and, ultimately, lost capacity.

In the office and warehouse, this can involve errors such as those made during data entry, those made in receiving, and picking and shipping the wrong product (or the right product, but to the wrong customer). This can be the result of a lack of standardized work and a lack of a visual workplace (to be discussed later, but having no checklists, forms, or directions), poor lighting, or lack of training.

As a consumer, how many times have you had an item you were purchasing at a retailer either not be in the system at checkout or instead have the wrong price rung up (usually higher)? According to an Federal Trade Commission (FTC) study, 2.24 percent of transactions result in overcharged items. This might not sound like a lot, but in fact, consumers lose $1 billion to $2.5 billion annually because of scanner errors. So, the fact is that shoppers do get overcharged and undercharged because of human errors, faulty UPC codes, and sloppy management practices.

This not only leads to frustration for the consumer (and the customers behind him or her at checkout), but in some cases results in either a lost sale or, in the case of grocery stores, a free giveaway to keep you happy.

There are many causes for defects and errors, such as poor processes, too much variation, supply issues, insufficient or improper training, tools and equipment not properly calibrated or precise, bad layouts, excessive or unnecessary handling, and inventory levels that are too high (i.e., inventory sits around longer, so there is more potential for damage).

Behavioral Waste (or Underutilized Employees)

Some people add an eighth waste, behavior. This is critical to consider, as you need employee creativity and participation to eliminate the other seven wastes. However, in some companies, there is a culture of not wanting to question things, not taking risks, or not rocking the boat. You might hear someone say, "This is how I was shown how to do it," or, "We've been doing things this way for years." If you're going to have a successful Lean journey, this type of behavior is unacceptable and must be changed.

A company's culture will be discussed later in more detail, but suffice it to say that a culture of team-based continuous improvement is a must. You must fully utilize and leverage employees' knowledge and skills, offer proper training, and provide opportunities for advancement to guarantee success.

Waste in Retail Store Operations

Once you've gained an understanding of the concept of waste, it's not so hard to find. In store operations, we can see it every day in the form of wrong pricing, incorrect planograms (a visual representations of a store's products or services), lack of category optimization, packaging, repackaging, back room push, back room storage, leftover seasonal stock, and the like. All of these types of waste can cause confusion and frustration to the customer, which is the last thing we want.

From the customer perspective, the shopping experience can be impaired by:

- Product proliferation and duplication
- Product complexity and change
- Confusing promotions
- Ineffective signage and excessive clutter
- Products hard to find or out-of-stock
- Poor service
- Long lines

This may result in:

- High levels of consumer stress
- Procrastination
- Reduced shopping frequency and duration
- Low purchase conversion rates

In Chap. 1, we discussed the five transition points in retail (i.e., enter, seek, find, select, and transact) and opportunities for process improvement. We can now link some of the Lean wastes just discussed with these transition points:

- *Enter*—Motion, waiting, and excess information
- *Seek*—Excess information and inventory, motion, and defects
- *Find*—Overprocessing, excess motion, waiting, and defects
- *Select*—Excess information, defects, transportation, and motion
- *Transact*—Defects or errors, waiting, motion, and overprocessing

Thinking Differently

As you can see, the concept of waste that is crucial to Lean thinking is not really that complex.

It is really just a different and easy way of looking at things. Once you start thinking this way, both as an individual and as a team, it becomes easier to see where waste lies in your business. The next thing to do is to take a step back and see where some of these wastes might exist in your operations.

Basic Lean Tools

The saying "you cannot build a house without a solid foundation" definitely applies when discussing Lean. In fact, there is something called the "House of Lean" (Fig. 5.2) that helps to illustrate this concept.

Although the importance of having a Lean culture as a key success factor has been discussed already (and will be discussed in more detail later), understanding how and when to use Lean tools will now be discussed.

In many cases, *value stream mapping* (VSM; see Fig. 5.3) is typically the next step taken by management after gaining a basic understanding of general Lean concepts. However, it is best to wait to address the details of VSM until later in the book, as you first need to understand the basic and advanced concepts and tools (in some detail) that can be used to deliver the opportunities for improvement identified in a value stream map.

Briefly, VSM is a mapping tool that gives a 10,000-foot-level view of a process. Typically, it is for a family of goods or services, starting from the customer and working its way upstream all the way to key suppliers. A VSM is similar to a flowchart or process flow map, but one of the key differences is that the "current-state" map identifies value-added and non-value-added activities. The "future-state"

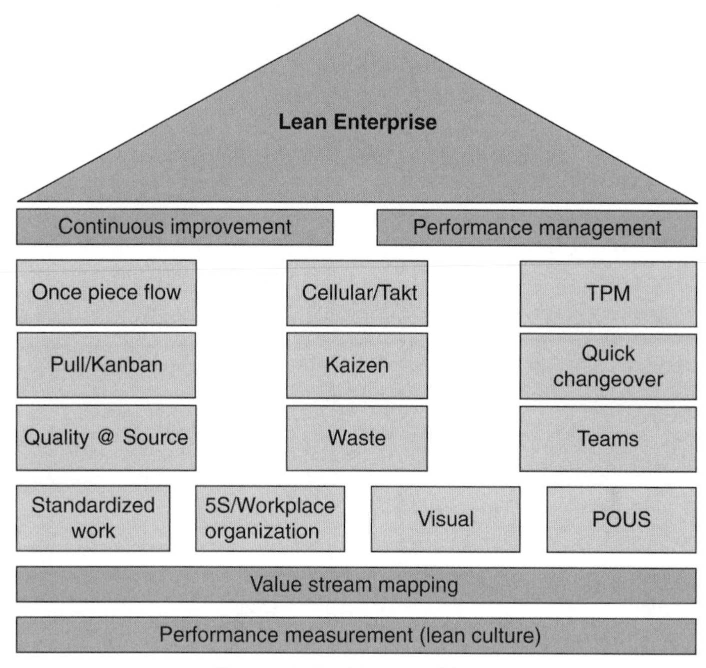

Figure 5.2 House of lean.

map, which can be thought of as a road map, attempts to reduce or eliminate the identified non-value-added activities. We will cover VSM in more detail in a later chapter on getting started.

Standardized Work

Standardized work refers to the standardization of best work practices—as the work is actually done in real life. The idea is to make work safe and repeatable with as little variation as possible along with high productivity. It is the best combination of employees, equipment, materials, and procedures.

There are examples of standardized work everywhere, including orders, drawings, *standard operating procedures* (SOPs), and the like. In fact, standardized work is one of the foundation principles of the TPS.

We know that in real life, while there may be SOPs in a binder on a shelf somewhere, most people do a job the way they were trained to do it (or, in many cases, the way they learned it on their own). Often this may not be the best way (i.e., in terms of method, sequence, or some other factor), but it is the way they were shown (e.g., "The guy I replaced showed me how to do this before he left") or have done it for many years (e.g., "I've been doing it this way for years, and it works for me"). The problem with this is that if everyone performs a task slightly differently, there may be variation, which can result in waste.

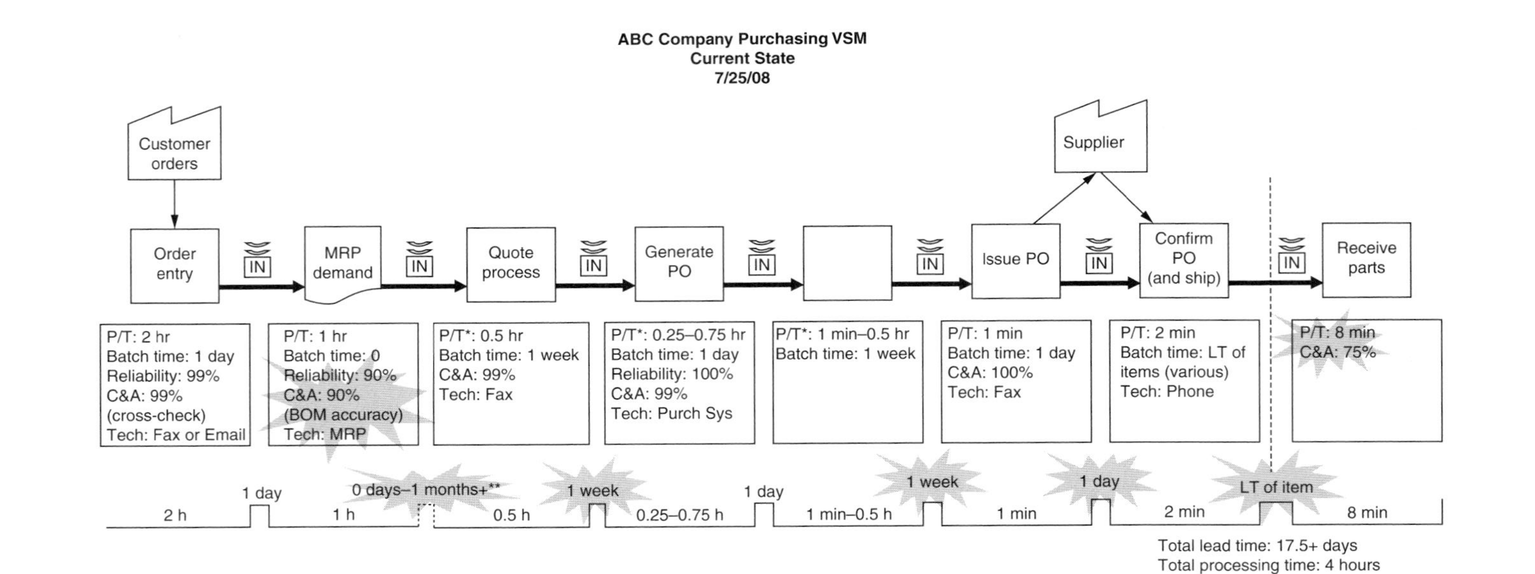

Figure 5.3 Value stream map.

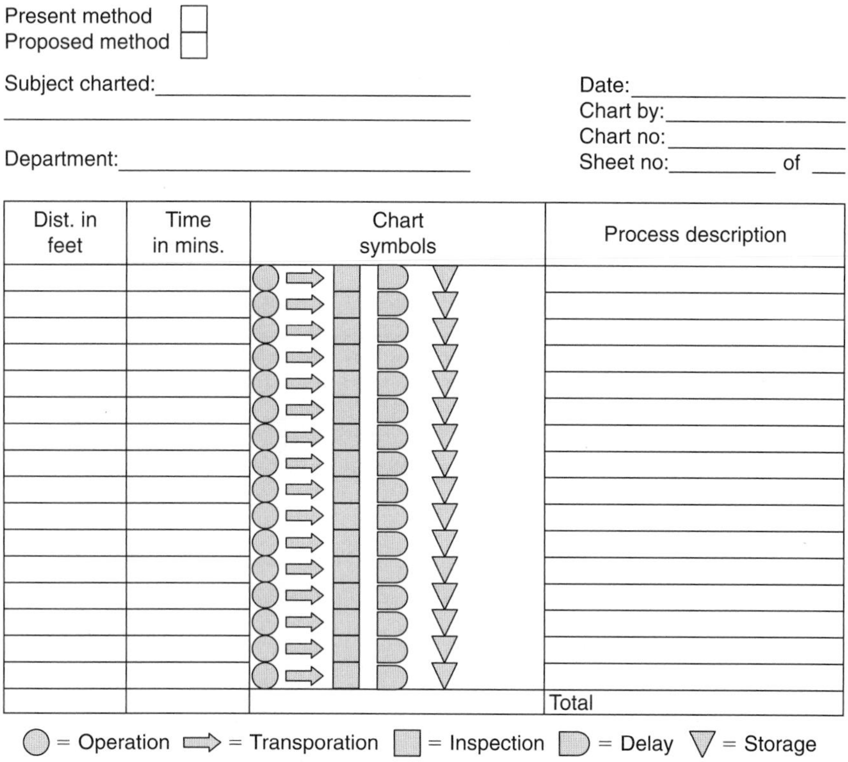

Present method ☐
Proposed method ☐

Subject charted:____________________ Date:__________
Chart by:__________
Chart no:__________
Department:____________________ Sheet no:______ of ___

Dist. in feet	Time in mins.	Chart symbols	Process description
			Total

= Operation = Transporation = Inspection = Delay = Storage

Figure 5.4 Process chart.

Usually it is best to get together a team composed of employees who actually do the work and coworkers from other areas, document the steps in the process (using digital photography), and come up with agreed-upon best practices, minimizing waste in the process. It can be useful to use a tool such as a *process chart* (see Fig. 5.4) to identify opportunities in the work process by capturing data for each activity, such as time and distance.

Visual Job Aids

It is then important to provide a *visual job aid* for this standardized work [see Fig. 5.5, a sample used at the DC of a national apparel retailer] that is easy to understand and follow.

This type of visual job aid should then be placed in the area where the work is done (laminated for protection) so that it can be followed by everyone. It is always a good idea to consider language restrictions in the workplace when creating standardized work (e.g., English and Spanish versions).

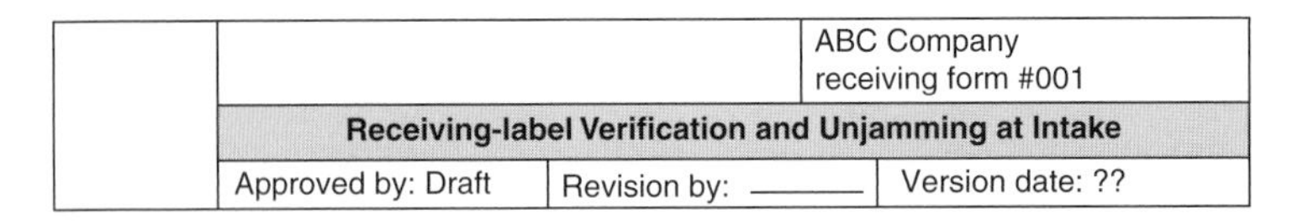

			ABC Company receiving form #001
	Receiving-label Verification and Unjamming at Intake		
	Approved by: Draft	Revision by: ______	Version date: ??

Phase	Step	Comments	Visuals
1. 1st electric Sys-label verification	• Inspect LPN and carton label • Press reset button	• Numbers must match • If numbers match, if not, take carton off conveyer and see supervisor	
2. 2nd electric Sys-un-jam (clear)	• Un-jam ("clear") carton • Push reset button		

Figure 5.5 Visual job aid.

Areas in retail and wholesale that this can apply to are found everywhere. The office, store, and warehouse or DC are the most common areas, and the work involved can include order processing, checkout, invoicing, and drawings.

In the warehouse, pretty much all the basic activities of receiving, putting away, picking, packing, and loading can benefit from visual job aids (see Fig. 5.6, a sample used at the DC of a national apparel retailer).

Visual job aids are especially important in warehouse and store operations, both in the office and on the floor, as there is a large use of temporary workers and relatively high turnover.

In warehousing, there are "lumpers" (outsourced workers who handle freight or cargo), temporary workers, and the use of *third-party logistics* organizations, or

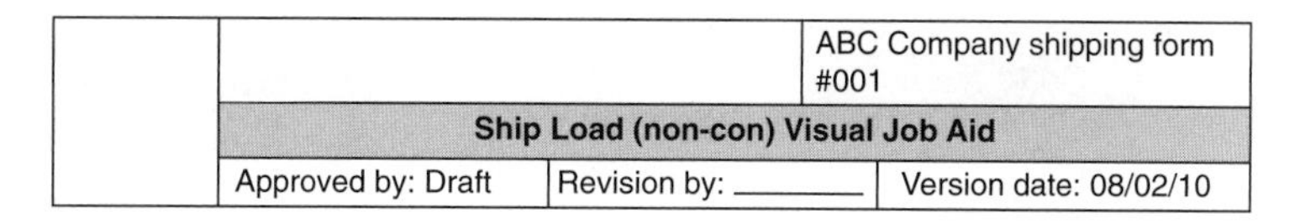

		ABC Company shipping form #001
Ship Load (non-con) Visual Job Aid		
Approved by: Draft	Revision by: ______	Version date: 08/02/10

Phase	Step	Checks/Remarks	Visuals
1. Move pallet(s) to door	• Move 1–2 pallets from door staging lane to door for loading	• # of pallets depends on equipment used.	
2. Scan pallet(s)	• Scan pallet label with RF gun	• Select option #5 on RF gun ("Load Carton/Pallet"); then enter Store # and Door #, then scan "Manifest Label"	
3. Load cartons into traller	• Load cartons into nose of traller if possible	• The objective is to pack cartons in tightly and securely as possible. • Large, heavy cartons should go on the floor with arrows up.	
4. Return empty pallet	• Return to designated pallet return area for that door.	• Located by door staging lane	

Figure 5.6 Visual job aid (loading a truck).

3PLs (i.e., outsourced logistics functions on a short-term or long-term basis), for which visual job aids are useful.

The same goes for store operations, where the use of temporary workers, especially during the holiday season, can lead to disorganization and confusion.

In an office, the longer it takes to process resupply information, the longer the "order-to-cash" cycle and the greater the chance for stockouts.

Standardized work then leads to an organized workplace that is neat, safe, and efficient, with a location for everything that is needed and the elimination of anything that is not needed in the area.

Visual Workplace

Think of the chaos that would result in our everyday lives if there were no speed limit signs and no lines on the highway or in a parking lot.

In manufacturing terms, there is something called a *visual factory*. These simple visual signals give operators the information that they need in order to

Figure 5.7 Simple kanban with visual signal example.

make the right decision. They are efficient, self-regulating, and worker-managed. Examples include visual job aids; signs; lines on the floor designating storage areas, aisles, work areas, and the like; "andon" lights (i.e., "and on" a red light, there is a problem, and the like); labels (color-coded in some cases); and *kanbans* (visual signals of the need to replenish inventory as a result of downstream demand; e.g., Fig. 5.7).

The visual workplace is one of the fundamental concepts of Lean. This translates easily to the warehouse. At first glance, a warehouse looks fairly organized, with bar codes and labels on pallet bins, safety lines on floors, and so on. However, a closer look usually reveals areas of clutter and disorganization (examples might include the supply and maintenance areas).

In retail, the back room can frequently be disorganized. Product availability and quick replenishment of the sales floor are critical to success in retail. The use of visual aids for everything from replenishment to a clean, organized work area can create huge benefits in terms of cost, efficiency, and safety.

When implementing visual systems, it is important that they are easily found where needed, are easy and quick to understand, and provide meaningful feedback.

Layout

Another key concept in Lean is flow. When we eliminate waste in a process, items keep flowing, as opposed to waiting in a queue, in an aisle, or somewhere else. Critical to this is the layout of a facility itself.

Typically, in manufacturing, companies grow "organically" and put things where they fit, not necessarily where they best belong. There tend to be "monuments," which are big, heavy pieces of equipment that are difficult to relocate. This may not be conducive to the continuous flow of materials or information.

In a warehouse, as travel time is critical to productivity, good layout (and flow) is essential. The idea of having your fast-moving A items located closer to shipping (and down low) and your slower-moving C items farther away and higher up (commonly called "velocity slotting") is not always carried out. Also, it is important to have tools, equipment, supplies, and packaging materials always available and close to where you need them.

The same goes for the office, where we tend to lose sight of how much walking we do in processing an order, for example. (It can add up to hundreds of miles per year of unnecessary walking!)

Retail store layout generates the overall look and feel of the interior of a retail store, including the positioning of fixtures (shelves, racks, and so on) and products within the store. This is an important part of implementing a retail store strategy. An effective layout will expose customers to as many products as possible within the amount of floor space that is available. This requires detailed marketing planning, ranging from deciding which products to sell and where to place them in the store, to how advertising materials will be displayed.

Let's look a little more closely at layout as it pertains specifically to retail.

Retail Layout

Walking Space

It is always a good idea for a store to allow adequate walking space for customers. Aisles should be wide enough to allow traffic to flow in both directions. When there are shopping carts involved, it is important that there be space for customers to stop and park without interrupting the flow of traffic. The requirement that customers be able to move freely up and down your aisles is important in retail layout design.

Flow

The layout must allow customers to enter from the front and walk to the back of the store. This is necessary in order to increase the amount of time that the customers spend in the store and increase the likelihood that they will buy more items. Always remember to disperse the high-selling items so as to enable customers to browse the entire store. Aisles should be in a horseshoe design when possible, as that moves customers through the front door, where impulse products and some high-demand items are located, and also gets them to the back of the store, where higher-priced items may be located.

Eye Level

Having items you are trying to move at the proper eye level will help to improve sales. At the same time, you must keep in mind whose eye level you are trying to reach (e.g., children or adults).

Display Cases

Display cases serve a number of critical functions in the layout of a retail store. Expensive products may be placed in a lighted display case to draw attention to them. The display case can also serve as a countertop customer interaction area for convenience, as the sales associate can quickly pull out an item from the display to show the customer. Displays can also be placed near the cash register area and, as a result, act as an impulse purchase area for customers. It can also be effective to put a display case at the checkout area with a salesperson there to interface with customers and answer questions [Walker, 2013].

In general, whether in a warehouse, an office, or a store, good layout results in:

- Higher utilization of space, equipment, and people
- Improved flow of information, materials, or people
- Improved employee morale
- Improved customer/client interface
- Flexibility

Workplace Organization and Standardization: 5S

Another tool, 5S, which stands for sort out, set in order, shine, standardize, and sustain, results in a well-organized workplace complete with visual controls, improved layout, and order. It creates an environment that has "a place for everything and everything in its place, when you need it."

The use of 5S produces a workplace that is clean, uncluttered, safe, and organized. People become empowered, engaged, and excited. A workplace that is clean, organized, orderly, safe, efficient, and pleasant results in:

- Fewer accidents
- Improved efficiency
- Reduced searching time
- Reduced contamination
- Visual workplace control
- A foundation for all other improvement activities

In the supply chain and logistics function, especially in the case of warehouse operations, this is often the first piece of Lean that is implemented. The main reasons are that it is a good foundation concept for future improvements, and that it is simple to understand and implement.

One of the leaders in this is Menlo Worldwide Logistics, a division of Con-way (www.con-way.com). It dedicates an entire section of its Web site to Lean logistics and states that "Menlo Worldwide Logistics practices lean logistics to deliver superior supply chain performance and give its customers a competitive advantage. Lean logistics emphasizes minimization of all resources used in supply chain management. The lean logistics methodology uses proven lean practices and principles to reduce waste, complexity and error. . . . Adherence to 5S leads to better quality service, lower costs, higher availability, higher customer satisfaction, and more reliable deliveries." [Menlo Worldwide Logistics, 2011]

Worker productivity in a warehouse or DC is especially critical (e.g., cases per hour, or CPH, is a typical productivity measure), and 5S can be especially useful in this regard.

As was mentioned previously, some large retailers, such as Wegmans and Sonae, have used 5S with great success in their store operations (especially in the back room/storage areas).

As was previously mentioned, the actual 5Ss are sort out, set in order, shine, standardize, and sustain. These will be defined shortly, but before starting on a 5S *kaizen* or *improvement* event, one must first perform a workplace scan. Typically, this entails the following steps:

1. *Area map*—Usually drawn by hand, it should show the area for which 5S is being done, including all machines and materials located in the area. Colored lines should show the movement of materials (and information) into and out of the area. This is called a "spaghetti" map for a reason—by the end, it usually looks like a bowl of spaghetti! It is a good way for the group to understand where there may be opportunities for improvement and where things are not flowing.
2. *A 5S audit*—There are many 5S audits available on the Internet. Basically, the facilitator leads the group through a "before" rating of the area in terms of each of the 5Ss (we will get there shortly), usually on a scale of 1 to 5, with 5 being the best. Typically, the first audit results in a fairly low score. The idea is that subsequent audits will find better results and can gradually be done less often.
3. *"Before" pictures*—It is always important to remember what the area looked like before 5S was carried out. The pictures can possibly be posted later on a 5S board with "after" pictures.

Once the team has selected an area and performed an initial workplace scan, the 5S process can start.

Sort Out

The first S is for *sort out*. This involves removing anything that is not needed in the area. Items can range from garbage, which can be disposed of immediately, to excess inventory, equipment, tools, furniture, and other things that don't belong there. As the saying goes, "When in doubt, toss it out."

Why do you need to do this? Throughput is increased as a result of improved workflow, communication between workers is improved, product quality is increased, wasted space is reduced, time spent looking for parts or tools is reduced, and overstocking is avoided.

Included in this phase is something called a *red tag* strategy. Simply put, a nearby area needs to be designated as the place where sorted-out items can be taken for future disposition. This is called a red tag area. Each item taken there should have a red tag attached with a variety of information, such as an assigned number, description, and recommended action (see Fig. 5.8).

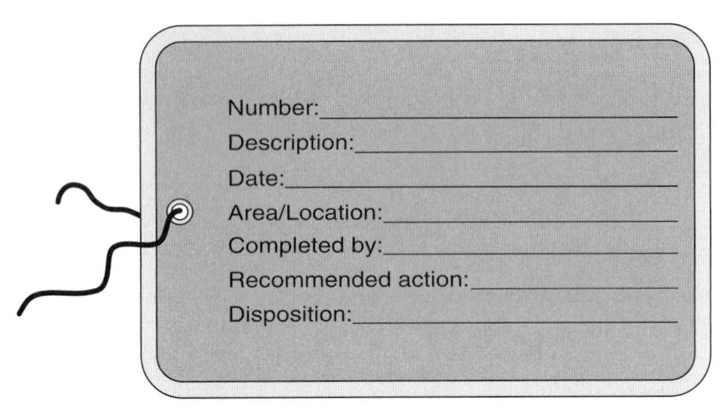

Figure 5.8 Sample red tag.

Each red tag item should be included on a corresponding red tag disposition sheet that is communicated to the appropriate parties, noting that red tag items must be claimed and removed within a certain time period (usually no more than two weeks), or they will be disposed of.

Surprisingly, there are quite a few categories of items and actions that can be taken during this process, as can be seen in the sample disposition list in Fig. 5.9.

Category	Action
Obsolete	• Sell • Hold for depreciation • Give away • Throw away
Defective	• Return to supplier • Recycle
Scrap	• Remove from area to proper location
Trash	• Throw away • Recycle
Unneeded in this area	• Remove from area to proper location
Used daily	• Carry with you • Keep at place of use
Used weekly	• Store in area
Used monthly or less	• Store where accessible in facility
Seldom used	• Store in distant place • Sell • Give away • Throw away
Use unknown	• Determine use • Remove from area to proper location

Figure 5.9 Disposition list.

Set in Order

After unnecessary items have been removed, everything that remains should be *set in order*. As the saying goes, "A place for everything and everything in its place." During this phase, a great deal of thought should be given to the area's layout and the flow of materials and information. This is a great time for the use of visuals as well. Color coding and outlining are a few ways to make it clear where things go. It is a good idea to have a label maker and masking tape to label shelves, bins, drawers, and other such things temporarily and to mark floors. A few weeks later, after people have "lived with it" for a while, everything will be made permanent.

A great visual to create in this phase is something called a *shadow board*. It can be as simple as taking an ordinary Peg-Board and outlining and describing the tools that go on it. That way, when workers see an empty spot on the board, they know that a tool is missing. It is also a good idea to label the tools hung on the board with the same information, so that when someone finds a tool, he or she knows where it goes.

Now is also the time to consider excess motion waste. You can also consider what was found in the area map done earlier.

Shine

The third S is for *shine*. During this phase, everything is cleaned and sometimes even painted. One of the key purposes of cleaning is to keep all equipment in top condition so that it is always ready to be used.

If the shine phase is not carried out, problems can come up, including poor employee morale, safety hazards, equipment breakdowns, and even possibly an increased number of product defects.

During the shine process, we clean away trash, filth, dust, and other foreign matter. Contamination can include debris, oil, documents, water, dirt and dust, food and drink, the result of poor work habits, and materials left by other people, such as maintenance.

As a group, you need to determine what needs to be cleaned, who is responsible for cleaning it, how the cleaning is done, and what tools are needed. Cleaning supplies should be neat, clean, well-organized, and readily available (see Fig. 5.10, a sample used at the DC of a national toy retailer). In many cases, areas have a "five-minute cleanup" at the end of a shift. Housekeeping checklists are always a good idea (like you see in McDonald's bathrooms showing when was the last time the bathroom was cleaned and by whom).

There is also the concept of "cleaning as a form of inspection," which involves keeping the workplace clean, inspecting while equipment while cleaning it, and, as a result, possibly finding minor problems during cleaning inspection. During the process, a greater emphasis is placed on the maintenance of machines and equipment.

The first three Ss produce the condition in which you want the workplace to be kept. In a way, they are like spring cleaning, except that, if they are implemented

Figure 5.10 Cleaning tools.

properly, you do not have to do them every spring, as the workplace is maintained at that level!

The next two phases, standardize and sustain, are all about *keeping* the workplace safe and organized.

Standardize

The fourth S is for *standardize*, which means creating a consistent way to carry out tasks and procedures—everyone does them the same (documented) way. It is really a form of the standardized work that was discussed earlier in this chapter. The idea is to standardize how the first three Ss are maintained.

For example, you may review the area weekly to see if anything needs to be red-tagged and removed (sort out), check inventory levels for supplies at the end of a shift (set in order), and have a five-minute cleanup at the end of the shift (shine). A good way to do this is to create a job cycle chart (Fig. 5.11), so that duties can be assigned and communicated.

Many companies employ the use of a 5S board for an area (or, in some cases, for a facility). The 5S board is an idea place where things such as current 5S audits, before and after pictures, housekeeping checklists, area maps, and job cycle charts can be displayed.

	5S Standardize: Job cycle chart		Name:________			Dept:________			Date:________		
No.	5S Job	Sort	Set	Shine	Standardize	Sustain	Continuously	Daily (AM)	Daily (PM)	Weekly	
1	Red tag (entire plant)										
2	Red tag (cell/line)										
3	Inventory check										
4	Tool check										
5	Wipe area										
6	Vacuum area										
7	Machine clean inspection										
8	Degrease work area										

Figure 5.11 Job cycle chart.

Sustain

The final S, and perhaps the hardest to accomplish, is *sustain*. Sustain refers to making a habit of maintaining correct procedures over the long term.

No matter how well we implement the first four Ss, improvement gains may be lost and 5S doomed to fail without a *commitment* from everyone (from management down to operators) to sustain it. Thus, 5S does not end at the conclusion of the 5S kaizen event. It must become part of a company's "culture" and become a habit if it is to be successful. There is no simple answer to being successful with 5S. It is a combination of communications, management support (including a Lean champion to spearhead the program and possibly a coordinator), culture, and rewards (everyone always wants to know, "What's in it for me?").

Communication methods can include 5S posters, before and after photo exhibits, 5S newsletters, 5S manuals (in English *and* Spanish!), 5S events, competitions, 5S department tours, and success stories. I cannot tell you how many times I have heard the story, "We did a 5S of our area and it worked fine for a while, but the second shift messed it up."

The use of 5S should become an everyday activity, with daily cleanup and weekly 5S activities. It should become part of everyone's job descriptions, and it should be measured and displayed on the 5S board.

While this chapter covered some basic tools for Lean, especially 5S, which is a great place to start in an office, warehouse, or store, the next chapter will get into some more advanced topics that can have an even greater impact on your company's bottom line.

CHAPTER 6

Thinking Lean: Advanced Lean Concepts and Tools—KISS (Keep It Simple and Straightforward)

The nice thing about Lean as a form of continuous improvement is that all the concepts and tools (even the more advanced ones covered in this chapter) are fairly easy to understand—it's not rocket science.

But it is important to understand that both the tools we covered in the last chapter and the more advanced ones we will discuss here are fairly easy to grasp. (Many of them are also used in Six Sigma, a more quantitative tool to eliminate variability in a specific process, and Lean Six Sigma, the recent merger of Lean and Six Sigma, as we will discuss later in this chapter.)

It is actually the ability for people to change and commit that is perhaps the hardest part, and we will talk more about that in Chap. 15, "Defining and Measuring Success," later in this book.

Batch Size Reduction and Quick Changeover

There are two critical concepts that go hand-in-hand in any Lean program. They are the ideas of batch size reduction and quick changeover (sometimes also referred to as setup reduction).

If you recall the difference between push and pull production, you can better understand this topic. In push, you produce in large quantities to spread your fixed costs over a large number of items, thus minimizing your costs per unit. In pull, you schedule closer to what the customer actually wants (i.e., you make what you sell). The ultimate goal is one-piece flow. While this may be unattainable, it is the direction that you want, and need, to head toward.

There are many benefits to this just-in-time (JIT) approach. In Fig. 6.1, which compares the two approaches, you will notice a few things. First, you can see that using smaller batches reduces the overall cycle time for any one item.

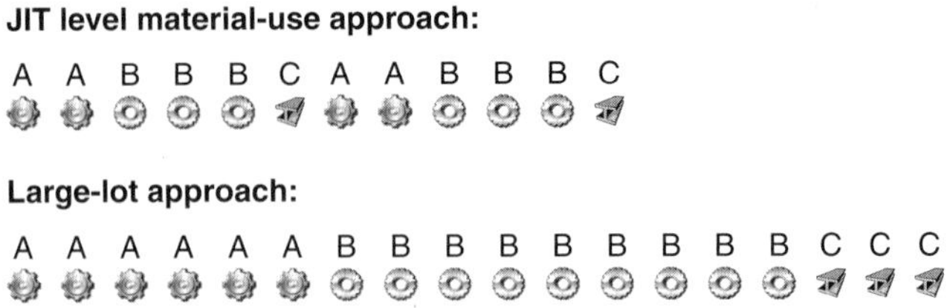

JIT level material-use approach:

A A B B B C A A B B B C

Large-lot approach:

A A A A A A B B B B B B B B B C C C

Figure 6.1 Push versus pull.

In the push approach, you have to wait for the large batches of other products scheduled to be completed before you even get to the one you are waiting for (item A in this example). You can also see that work-in-progress (WIP) is significantly reduced by the small lot approach. Finally, in the event that there is a quality problem that might affect the entire batch, it becomes less of a problem because of the batch size reduction.

Batch Size Reduction

The benefits of batch size reduction can include reduced lead times, lower inventory levels, more flexibility to meet fluctuating demand, better quality with reduced scrap and rework, less floor space used in production and storage, and thus lower overall costs.

In the supply chain and logistics area (not only in manufacturing, but also in wholesale and retail), there are the obvious results of batching in production and purchase orders to cover manufacturing wastes, resulting in excess inventory, and in purchasing to obtain economies of scale (i.e., to get better pricing, which will be discussed further in Chap. 8), but it can also be seen in the office, where batching typically occurs in the form of paper that can pile up in people's inboxes. There is a natural tendency to batch in an office, as there are usually some kind of setup for each type of activity, such as order processing, in which files, forms, faxes, and reference materials are gathered before going to a specific computer screen. Each step in the process is typically done in batches and therefore ends up sitting in someone's inbox until that person can get to it.

In retail, we order in batches not only from the vendor (i.e., the manufacturer or wholesaler), but also from our own distribution centers (DCs) in many cases. This replenishment ordering is also done in batches (albeit smaller than those from the manufacturer) and can thus be "pushed" downstream to our store back rooms and front areas, causing all kinds of waste.

Quick Changeover/Setup Reduction

The primary obstacle to reduced batch size is changeover time and costs. (Note: the typical definition of a changeover in manufacturing is "the time from the last good part to the first good part.") The goal is to minimize changeover time and cost so that smaller batches are run more frequently, resulting in better flow.

To go from one activity to another, whether on the shop floor or in the office or DC, requires some kind of changeover, which includes some kind of setup. As a result, batching seems like the most efficient way to do things (i.e., large batches mean fewer changeovers). Therefore, high setup costs encourage large lot sizes, and reducing setup costs reduces lot size and reduces average inventory.

If we can reduce the time and cost of making the changeover, batch size reduction can be realized. There is a concept called "single-minute exchange of dies" (SMED), which, while it literally applies to production operations involving a die, is used generally to refer to quick changeovers (or setup reduction) that result in smaller lot sizes and improved flow. Changeovers in manufacturing can vary in time from minutes to hours, but the idea is, through team-based continuous improvement, to keep reducing changeover time and cost so that things are produced in smaller batches or lot sizes. Think of a race car pit crew and all that it does in a very short amount of time (Fig. 6.2). While the car is being serviced, it is not in the race, so the more quickly the crew can get it back on the track, the better chance the driver has of winning the race. The same thing applies to changeovers and setups in business. If a piece of equipment is down, you are not able to make the product. The longer and more costly the changeover is, the less you want to do it, with the end result being large lot size batches. So if you can reduce the time and expense of setups, you can then change over more often, making smaller batches and getting closer to one-piece flow.

Typical changeover tasks can include preparation and adjustments, removing and mounting, measurements, settings and calibration, and trial runs and adjustments. If, as a team, you focus on waste in the current changeover process, it is not that hard to reduce the time it takes to make a changeover. Even a few minutes may be critical.

Most people who are involved in setups and changeovers feel that they are doing them in the best possible way. However, looking at the changeover in a kaizen event, with a mix of people from various functions, can shed new light on the subject. Typically, the team will observe and document the process (more than once if necessary) from beginning to end.

A useful tool in quick changeover is a setup analysis chart (Fig. 6.3). While a changeover is being observed (or videotaped, as it may take hours), every step in the process should be documented, including how long it takes, whether the step is

Figure 6.2 Car pit crew.

		Current method					Proposed method				
Seq. #	Element description	Internal	External	Duration	Distance traveled	Improving ideas	Internal	External	Duration	Distance traveled	Comments

Figure 6.3 Setup analysis chart.

"internal" (preparation while equipment is down) or "external" (while equipment is running), and distance traveled.

Some general keys to improving a setup are:

- ▲ Try to separate preparation or external setup from actual or internal setup and move as much as possible to external so that you can shorten the changeover time.
- ▲ Move the material and tools needed for the changeover closer to the actual spot where they are needed.
- ▲ Standardize the actual process (and combine steps where possible) and the tools used.
- ▲ Train operators and mechanics on procedures.

The net result should be an improved, shortened changeover.

In the supply chain and logistics area (wholesale and retail), as mentioned before, there is a large amount of batching of paperwork in an office, which if reduced can encourage improved flow and getting orders out faster, resulting in a shorter order-to-cash cycle.

In warehouse operations, there are setups everywhere, including receiving, picking, staging, loading, and shipping (especially for shift startups). Usually the first half-hour or more of a shift is fairly unproductive, so the more that can be standardized and visualized, the more productive personnel can be right from the start (especially temporary labor).

Figure 6.4 shows a receiving door that has been organized and visualized so that the operator can start up quickly, as everything that the operator needs to do the job is in place and ready to start unloading.

Figure 6.4 Organized receiving door.

Stores can have rough startups as well at the beginning of the day and when changing shifts, especially with the large use of temporary workers. So, standardizing and visualizing store openings and shift changes can make this activity more productive.

Additionally, many of the same issues with setups in a warehouse are evident in the receiving area in the back room of a store.

Kanbans

A key to successfully going from a push to a pull environment is the use of kanbans.

A pull system typically uses signals to request production and delivery from upstream stations (it might be a card with replenishment information, or as simple as a line on the wall; for a simple example, see Fig. 5.7). Upstream stations produce or replenish only when signaled.

The tool used to carry out this process is called a *kanban*. It is essentially a way to control the flow of materials and other resources by linking functions with visual controls. Only what has been consumed by the downstream customer is replaced. This determines the production and replenishment schedule.

By pulling material in small lots, inventory cushions are removed, exposing problems and emphasizing continual improvement. There are many benefits to a pull system, including reduced cycle time; fewer orders "dumped" downstream, creating excess WIP; less reliance on a forecast; and short lead times for customized products or services. The most common thing you may hear about kanbans that failed is that "we had one, but it stopped working." One of the main reasons for this is the establishment and maintenance of reorder points and reorder quantities. Lead time and demand rate to replenish can determine when to order, and the economic order quantity (EOQ) model, which minimizes inventory holding and ordering costs, can be a way of determining how much to order. In many cases, an item's demand may have peaks and valleys. As a result, it is necessary to continually review kanban order points and order sizes, as they may need to be adjusted based upon the current demand rate.

Many companies also try to use a "one size fits all" type of model when establishing kanban reorder points and quantities. This, of course, dooms the program to failure, as each item has its own individual demand characteristics and needs to be looked at individually.

In supply chain and logistics, kanbans have many applications, starting with the obvious one of replenishment of raw materials by vendors. This replenishment method can be initiated manually, with a kanban card being pulled when inventory is running low. The kanban card has basic item information, such as reorder quantity, pricing, and the like, needed to place a reorder manually or electronically based upon a predetermined reorder point. The extreme of this is called vendor-managed inventory (VMI), which will be discussed in more detail in Chap. 13, where the supplier takes full control over replenishing materials, saving the customer the costs of monitoring inventory and placing orders. Typically, either the vendor makes regular visits to check on inventory levels or the information is transmitted to the vendor electronically (there are now even vending-type machines to dispense and automatically reorder parts and hardware directly from the vendor). This results in orders being pulled in small lots, more frequently than was the case before the VMI (and also eliminates the need for you to check inventory and place purchase orders).

A kanban can also be used to replenish supplies and packaging materials. A good use for kanbans in DCs is to replenish supplies, such as labels, tape, and corrugated boxes, in an office or work area. It can be as simple as a line on a wall to determine the reorder point (see Fig. 6.5, a sample used at the DC of a national toy retailer).

Kanbans help to enable the concept of point of use storage (POUS), which is the idea of having the things you use more often closer to you (and typically at waist level), and the things you use less often further away and higher up. The kanban can be used to keep minimal amounts of raw materials and supplies nearby without taking up a lot of space in the work area, where space is at a premium and efficiency is everything. It also simplifies physical inventory tracking, storage, and handling and is a foolproof way to ensure that an area never runs out of needed materials or supplies.

A good example of a kanban in retail is the "partnership between JC Penney and TAL. JC Penney sends point of sale (POS) retail data directly to the factories, which enables factories to forecast demand, buy fabric just in time, and cut it to a particular stock keeping unit (SKU) or size immediately. Then, TAL speeds up the process even further by shipping the finished garments directly to JC Penney stores" [Kurt Salmon, 2013].

Figure 6.5 Simple kanban in a major retail DC.

Quality at the Source

Another Lean concept is quality at the source (also known as source control).

The idea here is that the next step in any process is the customer, and you want to make sure that you deliver perfect products to that customer.

A way to help ensure this is through the use of a *poka-yoke*, a Japanese term for mistake-proofing. A poka-yoke is a way to make it virtually impossible to pass on a defective part or piece of information from one process to another (think of the saying, "You can't put a square peg in a round hole").

An example of this at retail (actually two examples) would be the method used by McDonald's to serve french fries. McDonald's uses an aluminum scoop to position the fries vertically to a properly sized container, making sure that the portion served is correct.

Poka-yokes aren't limited to physical product, where you can create a foolproof device to ensure that the correct product is passed on (e.g., creating a right-angle "jig" to test the product before passing it on, similar to the McDonald's example). It can also be used for information by limiting the choices on a form or a screen, for example.

Quality at the source is typically used in conjunction with a total quality management (TQM) program. TQM is similar to Lean Enterprise in that it is a team-based program that spans the entire organization, from supplier to customer, and requires a commitment by management to have a long-term, companywide initiative to improve quality, as defined by the customer, in all aspects of products and services.

There are seven tools of TQM. They are continuous improvement, Six Sigma, employee empowerment, benchmarking, JIT, Taguchi concepts (specific statistical methods developed to improve quality), and knowledge of TQM tools such as Pareto charts and cause-and-effect or fishbone diagrams. Some of the seven tools of TQM are also found in Lean thinking (continuous improvement, empowerment, and JIT).

Work Cells

Another powerful tool that can have a significant impact on the efficiency of the workplace is a work cell. Work cells rearrange people and equipment that would typically be located in various departments into one group so that they can focus on making a single product, providing a single service, or providing a group of related items or services. It is not a new idea, as it was originated in 1925 by R. E. Flanders.

Work cells are typically laid out in a horseshoe shape, which allows for a more efficient use of work space and equipment and is conducive to one-piece flow and, as a result, less WIP. There is also the benefit of needing fewer workers, as each worker in the cell is able to perform all the activities required to produce or assemble the product or to deliver the service. Employees in a work cell typically have higher morale as a result of greater participation in the entire process.

The typical first step is to identify "families" of products or services (a family should have most, but not necessarily all, of the same steps). As a result of the job enlargement, a typical feature in a work cell in which employees have multiple responsibilities is that there is a great deal of training required, which results in a lot of flexibility within the cell. There is often an opportunity to use poka-yokes to assure good quality as well.

Balancing a Work Cell

It is then important to calculate the takt time for the product or service family (total work time available divided by the number of units required) in order to balance the activities in the work cell so that materials or information can flow. So, for example, if the daily demand for a product (on an eight-hour shift, assuming no breaks in this example) is 800 units, the takt time is 100 units/hour, or 1 unit every 36 seconds. Each activity in the cell should be capable of making and passing 1 unit every 36 seconds in order for the cell to be balanced. This information can help to identify bottlenecks in the process, where one or more components or resources limit the capacity of an entire system, so as to make sure that the cell is balanced properly (i.e., that each step in the cell is capable of processing one unit at the required rate of takt time).

In the event that the cell is not balanced, having flexible, cross-trained employees can help address bottlenecks; in the case of machine bottlenecks, other approaches may be necessary, such as running overtime, speeding up, or adding equipment.

A time observation form (Fig. 6.6) is a useful tool when you are looking to staff and balance any multiple-step process. You take multiple measurements of the time required to perform each activity to see whether the entire process is balanced and whether there are any bottlenecks (relative to the takt time). It can also help to see whether you have the proper number of people involved in the process (total observed time to complete all activities for one part divided by takt time).

Work cells can be appropriate not only on the shop floor, but in the office and the warehouse. For example, in a distributor's office, there are various functions that are required in order processing, including receive order, check credit, review and enter order, reconcile and confirm order, and finalize and release order. Typically, these activities are done by different people, who may even be in different departments. Performing these functions can take a day or longer, while the actual value-added activities may take only 30 minutes or so. Along the way, there may be batching of orders, waiting for approvals, and a lot of walking around.

If this function were to be set up in one work cell where all of the activities were performed, the entire process could take less than 30 minutes for each order using fewer employees (who have had their jobs expanded). The end result would be that the orders were released for picking and shipment much more quickly, thus speeding up the order-to-cash cycle for the business. If you are implementing a work cell, you should realize that higher wages and more training may be required as a result of more responsibilities, but this can be well worth it overall.

In the DC itself, there may be more limited opportunities, but they are typically found in areas like packaging or value-added activities performed by third-party logistics organizations (3PLs), such as packaging of kits for a customer.

Use this form to help design a specific workstation within a cell

Step 1: Down the left side of the form, list the component tasks in the order in which they are performed.
Step 2: Observe each component task multiple times.
Step 3: Average the times for each component and the total cycle times.
Step 4: Estimate the value-added and non-value-added time for each component task.

Time Observation Form												Date:		Event or observer:	
												Model:		Takt time:	
#	Component task	1	2	3	4	5	6	7	8	9	10	Avg. time	Value-added time	Non-value-added time	Comments and observations

Figure 6.6 Time observation form.

Many chain restaurants have knowingly (or perhaps unknowingly) created work cells for "assembling" meals. If you observe how compactly the kitchen area at a McDonald's, Taco Bell, or Subway is laid out, you will see that it is actually a work cell for assembling your meal. These chains are very flexible in that they can produce a wide variety of items in a short amount of time, in a small area, and by a small number of cross-trained employees. Small quantities of materials are typically located within arm's reach, with backups of the same size ready for replenishment (almost kanban in nature). Highly utilized, general-purpose equipment (e.g., microwaves and deep fryers) is also located nearby.

Total Productive Maintenance

The final major Lean concept that I would like to cover is that of total productive maintenance (TPM), which focuses on equipment-related waste. Equipment maintenance (or lack thereof) is often an overlooked area of waste. In fact, studies have shown that most manufacturers (70 percent or so) operate using what is commonly called *breakdown maintenance*. You would not treat your car that way. Every 3,000 miles or so, you take your car in to change the oil and air filters, lubricate various parts, check fluid levels, and the like. This is called *preventive maintenance* (PM). You do not wait until the transmission drops out on the road before checking the fluid, gears, and other systems.

There is another type of maintenance called *predictive maintenance*, but that is typically used after a good PM program is in place. In predictive maintenance, tools are used to check temperature, vibrations, and other such measures to see if some corrective action is necessary.

TPM, in and of itself, is not a PM program. It can, however, result in putting such a program in place.

Overall Equipment Effectiveness

Basically, a piece of equipment is observed (and possibly videotaped) during an entire shift to come up with an overall equipment effectiveness, or OEE, percentage (Fig. 6.7). According to various studies, typical companies average in the 70 percent area. That means that there is room in most companies to reduce or eliminate equipment-related waste to increase throughput and quality.

During the shift, observations of how the equipment is running are made. The observations are broken into three categories: performance efficiency, availability,

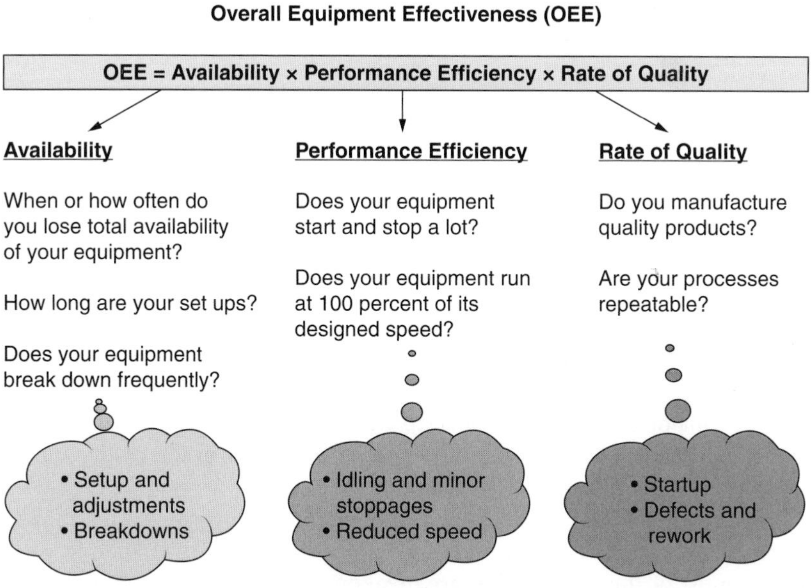

Overall Equipment Effectiveness (OEE)

OEE = Availability × Performance Efficiency × Rate of Quality

Availability

When or how often do you lose total availability of your equipment?

How long are your set ups?

Does your equipment break down frequently?

- Setup and adjustments
- Breakdowns

Performance Efficiency

Does your equipment start and stop a lot?

Does your equipment run at 100 percent of its designed speed?

- Idling and minor stoppages
- Reduced speed

Rate of Quality

Do you manufacture quality products?

Are your processes repeatable?

- Startup
- Defects and rework

Figure 6.7 OEE calculation.

Date:________________ Observer:____________

Equipment #________ Description:______________ Dept:________________

Start from	End to	Performance efficiency					Availability				Quality		
		Idling and minor stoppages					Breakdowns						
To	Running time	Chips	Jam	Insert	Other	Reduced speed	Lube		Other	Setup and Adj	Startup	Defects and rework	Comments

Figure 6.8 OEE observation form.

and quality. Within each category are specific reasons for slowdowns, stoppages, breakdowns, and quality issues. This information is put on an OEE observation form (Fig. 6.8).

After the shift is over, the various observations are tabulated and separated into three major categories to come up with an overall OEE percentage for the piece of equipment observed. As with a changeover kaizen event, a video is helpful in TPM to see more detail.

An analysis of the major losses (Fig. 6.9) is then performed. This shows the most productive place to begin an improvement process. In this example, you can see that most of the major losses are breakdowns, thus narrowing down where

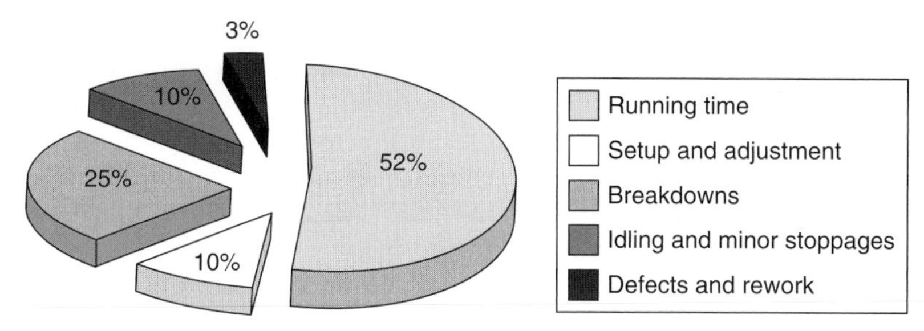

Figure 6.9 Analysis of major losses.

you should look to improve equipment performance. Once you have determined the major source of waste or losses, you can use a Pareto analysis to get to the root cause of the waste (Fig. 6.10). In this example, more than half of the breakdowns are in the load arm. By dealing with this one issue, you can make major progress.

The team, which should be made up of a variety of people, including maintenance personnel, engineers, and leads, should meet to discuss the major causes of waste and how to reduce or eliminate them. Solutions can range from simple, inexpensive solutions called countermeasures (e.g., putting a filter over a motor to reduce the buildup of dirt and grease within it) to daily, weekly, and monthly PMs (some performed by the operator and others by mechanics).

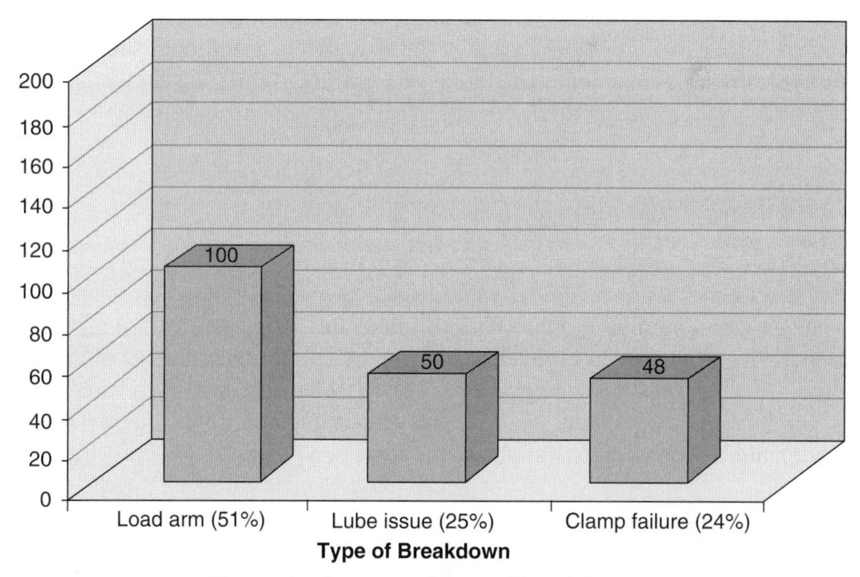

Figure 6.10 Pareto chart of breakdowns.

All of these solutions should be put into an improvement plan that describes the problem you are going to improve, a list of questions you have, a summary of action steps, and a plan to monitor improvements.

TPM can also be applied in the office, as there is plenty of technology that can affect productivity, such as computers, copiers, and fax machines, to name a few. The same thing goes in transportation and distribution, where things such as carousels, forklift trucks, automated storage and retrieval systems (ASRS), pallet wrappers, and radio-frequency (RF) devices can affect productivity.

Retail operations have plenty of equipment that needs to be maintained properly, ranging from the simple (shopping carts and racks) to the complex (bar code scanners and self-checkout equipment). If they fail, it can lead to customer frustration and stress, and possibly to lost sales.

Lean Analytical Tools

There are many analytical tools that can be used in Lean, some of which are also used in TQM, as mentioned earlier. They can best be divided into categories of tools for (1) gathering data, (2) organizing data, and (3) identifying problems.

Tools for Gathering Data

The tools for gathering data include simple check sheets, scatter diagrams, and cause-and-effect (or fishbone) diagrams.

Check sheets are simple "chicken scratches" to organize data; scatter charts are basically a graphical view of the relationship between two variables; and fishbone diagrams (so called because they are shaped like a fishbone) show process elements or causes of an outcome or effect. These are useful tools for teams to use to brainstorm improvement ideas by first trying to come up with possible reasons for waste.

Tools for Organizing Data

The tools used for organizing data include Pareto charts (Fig. 6.10) and flowcharts.

Pareto charts use the 80/20 rule (discussed previously), which states that a relatively small number of items typically generates a large percentage of sales or profits. This can apply to problem-solving as well, as a small number of types of issues typically generates a large proportion of problems or waste. So a Pareto chart or graph identifies problems or defects in descending order of frequency. On the other hand, a flowchart visually shows the steps in a process.

Tools for Identifying Problems

Tools for identifying problems may include the five whys, histograms, and statistical process control (SPC).

The five whys is a method in which you ask a series of questions in order to get to the root cause of a problem or defect. The idea is that the answer to each question leads you to ask another "why" until you get to the source of the problem.

Histograms are a graphical way to show the distribution of the frequency of occurrences of variables, such as problems or defects.

SPC is a huge subject, of course, and it is one of the major tools used in Six Sigma to minimize variability in a process. In general, SPC utilizes both statistics and control charts (which indicate whether a process is in a state of statistical control) to tell you when you need to take corrective action. It can be used for process improvement as well.

There are four major steps in SPC. First, if you don't measure a process, you can't control and improve it. A control chart has upper and lower control limits (UCL and LCL). If samples are within those bounds, then the process is viewed as being "in control." Second, before making a change to a process, you should make sure that you have an assignable cause. Third, you should try to reduce or eliminate the cause of the problem or defect; and fourth, you should restart the revised, improved process.

In terms of supply chain and logistics, many of the Lean Six Sigma tools mentioned earlier can be used to identify, analyze, and minimize variation in areas such as inventory control, forecasting and customer demand volatility, and on-time delivery, to name a few.

For example, a client of mine manufactures helicopter accessories. The team identified waste in the receiving area that slowed down the receiving and put-away process. It determined that the waste had three major causes: suppliers, resources, and workers. To both validate this and gather data, the team members had receiving personnel collect data for a month. They then used the Pareto principle to segment the major wastes in each of these categories, and then used a kind of informal "fishbone" to determine the major causes of the largest source of waste within each category (see Fig. 6.11). In the example shown, the data indicated that the largest supplier-caused waste was missing paperwork. The team then drilled into the collected data to determine that the major causes of missing paperwork were missing certification (58 percent of the occurrences) and missing blue/yellow tags (20 percent). Subsequent next steps and action items were taken to work with suppliers and internally to eliminate the two largest causes of missing paperwork, which made up 78 percent of the identified waste—a real improvement.

Lean Six Sigma

Six Sigma is a measure of quality that strives for near perfection and tries to improve the quality of the outputs of a process by identifying and removing the causes of defects or errors by minimizing variability in an activity or business process. It uses a set of quality management methods, incorporating statistical methods (including the ones mentioned earlier), and utilizing people within the organization (Champions, Black Belts, Green Belts, and so on) who are knowledgeable about the various

ABC Company
Receiving Kaizen Event–Supplier

% of Waste	Description
80%	Missing paperwork
15%	Incorrect/Incomplete paperwork
5%	Early delivery

% of Cause for Waste	Types of Missing Paperwork	Comments
58%	Certification	
20%	Blue/Yellow Tag	
15%	Age control	Biggest waste of time, but fewer occurances
1%	Check sheet	
1%	Packing slip	

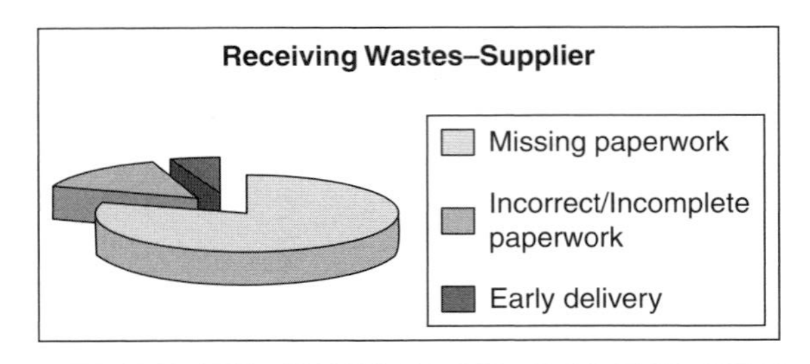

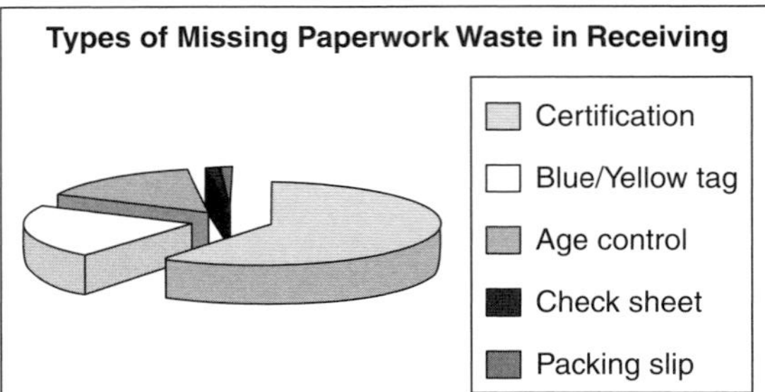

Summary of Next Steps:

(1) Receiving (John Smith and Joe Jones) will track occurances of each type of missing paperwork for 30 days (including type of missing paperwork, supplier, etc).

(2) John and Joe will analyze findings and take corrective action for major causes of waste.

Summary of Findings to Date:

(1) Review of one month's tracking found that main causes of missing paperwork were:

- Track card not signed/missing
- Ship to WIP not transacted
- Wrong Rec case (buyer not updating)

Also:

- Materials not issued
- Overshipped

Figure 6.11 Receiving kaizen event example.

concepts and tools utilized. Six Sigma projects follow a specific sequence of steps and have quantified value targets, such as cycle time reduction, increased customer satisfaction, cost reduction, or higher profits, and can result in a fairly quick payback.

Lean Six Sigma therefore combines the power of Six Sigma (strategic alignment, customer focus, and rigorous analytical tools such as those just described) with Lean's longer-term focus on waste elimination from the customer's perspective in a value stream. Ultimately, they are complementary in nature, and the sum is greater than the parts in many respects.

Lean Six Sigma in Retail

> Six Sigma is finding new life, especially in retail. Target (TGT) claims more than $100 million in savings over the past six years from the program.
>
> Mike Fisher, Best Buy's (BBY) senior director of Lean Six Sigma, says projects like streamlining appliance installation have helped the company save up to $20 million in some cases. "Without a doubt, it put us in a better position to muscle through the recession by getting all of those inefficiencies out," says Fisher. SSA has helped several retailers expand programs in the past year. Retail practice leader Suzanne Long says that three national retail chains are among the consultancy's new clients.
>
> For a sense of the new approach to Six Sigma, consider Target. It hasn't enforced its program with the rigor common in manufacturing. "Some companies require their employees to use Six Sigma," says spokesperson Beth Hanson. "You don't have to be a black belt or green belt or be certified at all to use Six Sigma [here]. We just offer employees the tools." [Burnsed and Thornton, 2009]

DMAIC

Six Sigma encompasses the Define, Measure, Analyze, Improve, Control (DMAIC) methodology for process improvement and focuses on the reduction of variability in an individual activity.

In general, by using many of the tools described in this chapter, Lean Six Sigma teams (and teams in general) can follow five problem-solving steps to quickly identify the root causes of problems, develop solutions, and put in place procedures that maintain those solutions. A formal Six Sigma model used for process improvement is known as DMAIC, which stands for:

- *Define*—Identify the customer requirements, clarify the problem, and set goals.
- *Measure*—Select what needs to be measured, identify information sources, and gather data.
- *Analyze*—Develop hypotheses and identify the key variables and root causes.
- Improve—Generate solutions and put them into action, either by modifying existing processes or by developing new ones. Quantify costs and benefits.
- *Control*—Develop monitoring processes to ensure continued high-quality performance.

The limitation of programs such as Six Sigma, especially in retail, is that they're typically implemented through a highly structured program staffed by experts (e.g., Six Sigma Black Belts), and in many cases, management doesn't stand fully behind such programs.

I believe that process improvement programs like Lean and even Lean Six Sigma have a better chance for success if we can keep things simple and make the program easier to implement in all areas of business, with all employees participating at least to some degree. That is what we are trying to accomplish with the tools and methodology described in this book.

All of the tools and methodologies discussed in this chapter are very useful for getting to the root cause of problems and defects, but in order to make them work for you, there has to be a team-based continuous improvement culture in place, which is described later in this book.

CHAPTER 7

Being Lean: Retail Strategy—Sales and Marketing, Location, Human Resources Management, IT, Supply Chain Management, and Customer Relationship Management

A retail strategy is essentially a plan that describes how a business intends to offer its products or services to consumers and influence their purchases. The strategy outlines the business's mission, goals, consumer market, both general and specific activities, and necessary control mechanisms.

This strategy ultimately drives all of the basic functional strategies, including sales and marketing, location, human resources (HR) management, information technology (IT), supply chain management, and customer relationship management (CRM), which will be covered in this chapter (more tactical decisions concerning merchandise management and operations will be covered in the next several chapters). There will also be a discussion of how Lean tools and methodologies can be incorporated into each of these strategies. Ultimately, these functional strategies, when successful, help a retailer to reach its financial objectives in terms of sales and profitability.

As today's economy is more challenging than ever, with new competitors, formats, technologies, and globalization, the strategy developed needs to deal with this environment in terms of the firm's customers, competitors, and partners if it is to be successful.

Retail Strategic Plan

The process of strategy formulation in retail (and wholesale, for that matter) is the same as that for any other industry. The general steps involved in formulating a strategic plan are:

1. Mission statement
2. Situation analysis
3. Setting objectives
4. Obtaining and allocating resources
5. Developing the strategic plan
6. Strategy implementation and control

The next section of this chapter will define and describe each of these steps when formulating a strategic plan for retail.

Mission Statement

The strategic planning process starts with the retailer defining or stating the organization's mission. The mission is at the core of the retailer's existence, while other aspects of the strategy may change over a period of time or vary for different markets.

The mission statement identifies the long-term purpose of the organization. It describes what the retailer wishes to accomplish in the markets in which it chooses to operate. The retailer's mission statement would normally highlight the following:

- The products and services that will be offered
- The customers who will be served
- The geographic areas in which the organization chooses to operate and the manner in which the firm intends to compete

Situation or SWOT Analysis

Once the retail mission is defined, the organization needs to look inward and perform a situation analysis. It needs to understand what its strengths and weaknesses are in order to look outward to analyze its external opportunities and threats. This process is known as a *SWOT analysis.*

The SWOT analysis helps the retailer determine its position and its strengths and weaknesses and helps it formulate a clear picture of the advantages and opportunities that can be exploited and the weaknesses that need to be worked upon. This forms the basis or the core element of any strategy.

After determining its strengths and weaknesses, the retailer needs to consider various alternatives available to tap a particular market. A common tool for this (Fig. 7.1), which presented a matrix to look at growth opportunities, was developed

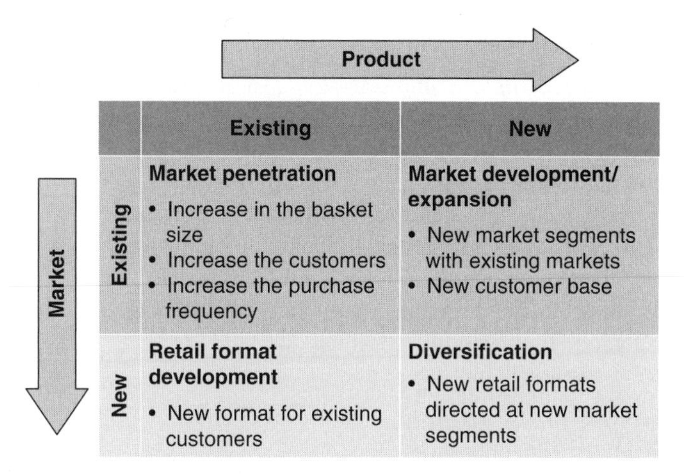

Figure 7.1 Ansoff matrix.

by Igor Ansoff in 1957 [Ansoff, 1957]. Ansoff focused on the firm's present and potential products in its existing and new markets.

Ansoff's matrix can also help us to understand the options that are available to a retailer. The alternatives available to a retailer under this model are the following.

Market Penetration

This strategy involves increasing the sales of an existing product and penetrating the market further by promoting the product heavily or reducing prices to increase sales; it is often used by supermarkets and large retail chains. This strategy may focus on any or all of the following:

- Increasing the number of customers
- Increasing the quantity purchased by customers (basket size)
- Increasing the frequency of purchase

Increasing the number of customers can be achieved by adding new stores and by modifying the product mix.

Another approach to increasing penetration is to encourage salespeople to cross-sell (i.e., to sell an additional product or service to an existing customer).

Market penetration strategy is the least risky strategy, as it leverages many of the firm's resources and capabilities. However, market penetration has limits, as once the market approaches saturation, a new strategy will need to be pursued if the firm is to continue growing.

Market Development/Expansion

This strategy is used when a retailer wants to reach out to new market segments or completely change its customer base.

The strategy involves:

- *Entering new geographical markets*—Expansion by adding new retail stores to an existing network is an example of geographical expansion.
- *Introducing new products to the existing range that appeal to a wider audience*—Introducing a office supply aisle in a supermarket (e.g., a Staples business supply section at a Stop & Shop supermarket) is an example of a retailer introducing new products that appeal to a different audience. Another example is McDonald's when it introduced high-end coffee service. This not only created add-on sales, but also brought in customers who had never thought of going to McDonald's for premium coffee.

Retail Format Development

Retail format development occurs when a retailer introduces a new retail format to customers. An example is when a fast-food retailer like McDonald's or Subway offers limited menus at small facilities located inside large department stores.

Another example of this is when the toy store retail chain Toys "R"Us opens smaller-format "pop-up" stores in malls around the holidays.

This strategy may be appropriate if the retailer's strengths are related to specific customers, rather than to specific products, as in this case, a retailer can leverage its strengths by developing and offering a new product targeted to its existing customers.

Diversification

Diversification involves selling new products to new markets. For example, if a retailer that usually sells groceries to families decides that it would like to sell high-end electronics to single men, this would be diversifying. Diversification is a high-risk strategy, as the firm is unfamiliar with both the product and the target market. However, as it also has the potential to produce the highest rewards, many businesses are prepared to take the risk.

Setting Objectives

This phase of strategic planning involves the translation of the mission statement into operational terms. You need to indicate the results to be achieved, give direction to the performance desired, and set standards for the measurement of that performance.

Management also needs to set long-term and short-term objectives. A one- or two-year time frame for achieving specific targets is typical of short-term objectives, whereas long-term objectives are less specific and reflect the strategic dimension of the firm.

When setting objectives, there are two important focus areas for retailers to keep in mind: market and financial performance. These objectives include targets for the following objectives:

- Sales volume
- Market share
- Profitability
- Liquidity
- Returns on investment

Obtaining and Allocating Resources

A retailer needs to consider both people and financial resource requirements.

When the firm is thinking about HR needs, the plan must be consistent with the organization's overall strategy, and HR management must focus on issues such as recruiting, hiring, training, compensating, and motivating personnel. These activities must be managed effectively and efficiently.

Financial resources take care of the monetary aspects of business, such as rent, salaries, and payments for merchandise. Many businesses fail as a result of their owners not doing a good job at projecting the financial resources needed to open and operate the business. I think we've all seen local retail businesses that seem to have spent a lot of money getting a business ready to launch, only to have it fail within six months. The thought that comes to mind is, "How could someone put all of those resources into the startup of the business and not have enough working capital to keep it going for at least a year?" Well, that is an example (at least in my mind) of inadequate financial resource planning.

Developing the Strategic Plan

At this stage, a strategy is determined, through which the retailer will achieve its objectives. The retailer determines and defines its target market (i.e., the segment of the consumer market that the retailer plans to serve), and then finalizes the retail mix that will serve the target audience.

While there is no one process for deciding on and selecting the target market, most retailers look at the entire market in terms of both size and the consumer segments to which it might appeal. From these segments, they identify a smaller number of segments that appear promising, which become possible targets.

Variables like growth potential, the investment necessary to compete, the strength of competition, and the like are evaluated. This enables the retailer to arrive at the best alternative that is most compatible with the organization's resources and skills.

Segmentation

Considerations for successful market segmentation include the following:

- *Measurable*—Is the segment measurable and identifiable?
- *Accessible*—Focusing the firm's marketing efforts on a particular market segment should have a positive impact toward gaining the desired response.
- *Economically viable*—The expense and difficulty of focusing the marketing efforts on potential segments should be justified.
- *Stable*—The consumer characteristics are indicators of market potential, so the stability of indicators should be considered.

Retail Mix and Positioning Strategy

After the target market is chosen, the retail mix needs to be developed. This process involves:

- Determining the merchandise mix.
- Selecting a pricing policy.
- Identifying the types of locations where the retail stores should be located.
- Determining the services to be offered.
- Deciding which communication platform the retailer should adopt.

After the retail mix has been developed, a positioning strategy needs to be formulated. This refers to the image the retailer wants customers and competitors to have in their minds about the products and services that it offers. For example, do we want to position ourselves as more of a mass merchandiser offering a broad variety of items, or more of a "niche" retailer offering a narrower, but deeper selection (e.g., Target versus Toys "R" Us).

Strategy Implementation and Control

Implementation is the key to the success of any strategy. The effective implementation of the retailer's desired positioning requires the following:

- Every aspect of the stores must be focused on the target market.
- Merchandising must be single-minded.
- Displays must appeal to the target market.
- Advertising must talk to the target market.
- Personnel must have empathy for the target market.
- Customer service must be designed with the target customer in mind.

After implementation, management needs feedback and should focus on the performance and effectiveness of the long-term strategy by periodic evaluation. This ensures that the plans do not become fragmented or have suboptimized efforts, and that all efforts are in harmony with the overall competitive strategy of the business. Management can also use the process to decide on any future policy changes that would ensure that the combination of the retailing mix variables supports the firm's strategy.

Lean Strategic Focus: Retail Value Chain

As we know, retail is very challenging and dynamic. Most retailers grow from a single shop to a chain of retail stores, then possibly from a local to a regional or even national presence. So strategy and planning become very important, and, as a result, the retailer should have a clear focus and strategy for long-term growth and profitability.

The value chain described in Chap. 2 (Fig. 7.2) [Porter, 1998] can help us set our sights on what adds value to the customer, as it identifies the critical elements in the value chain (i.e., inbound logistics, operations, outbound logistics, marketing and sales, service, procurement, technology development, and HR management).

A retailer can obtain a cost advantage by making individual parts of the value chain more efficient or by reconfiguring the entire value chain to its competitive advantage. This can include structural changes such as new production processes, new distribution channels, or a different sales approach.

Ultimately, these areas can become strategic "bases" of sustainable competitive advantage, as we then strategically focus our Lean efforts on more effective and efficient sales and marketing, location, HR management, information systems, supply chain and logistics management, and CRM, which we will now discuss.

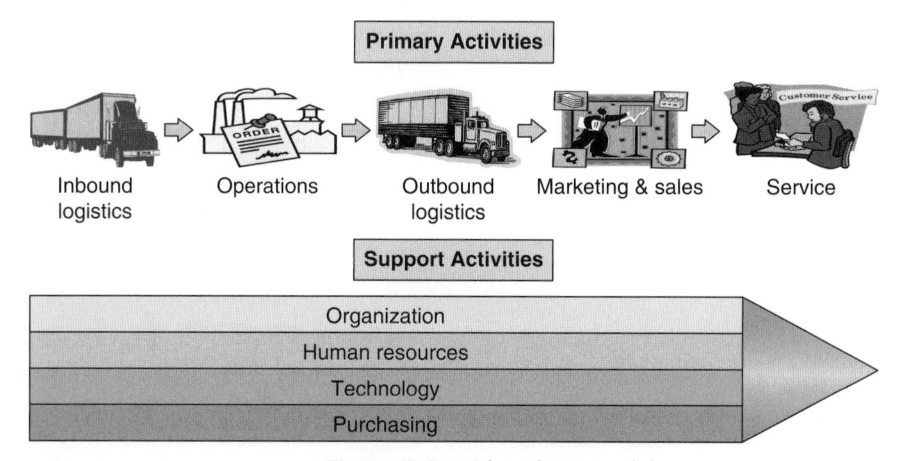

Figure 7.2 Value chain model.

Marketing and Sales Strategy

Retail marketing involves creating awareness and sales of the company's products. Unlike other types of consumer product marketing by manufacturers, retail marketing involves selling finished goods in small quantities to the consumer or end user, usually from a fixed location. Nonetheless, it still uses the common principles of the marketing mix: product, price, place, and promotion, which we will now discuss (see Fig. 7.3).

Product

Product marketing decisions vary based on the type of retail outlet for the target audience. Things that affect the consumer's perception of quality include packaging design, service plan options, warranty, colors, and materials.

For example, if you are marketing your retail store and its products to a high-end customer, you might choose an elegant design and cool colors; for a family-oriented store, you might lean more toward bright and bold colors.

Price

Price is, of course, important, as your customers must be able to afford your products or they won't shop at your store. So, the majority of the products should be affordable to your target audience base. A great deal of data has to be gathered and analyzed to determine your target audience's median income and spending power.

Place

The way a store is designed and how you display your products is a critical part of the marketing mix. The display has to fit your desired image, so that customers have a positive, convenient experience. In a luxury retailer, products may be placed father apart to create a sense of scarcity and exclusivity; in an electronics

Figure 7.3 Marketing mix.

store, display models are set within arm's reach to allow for close examination and possibly testing. Placement can be useful in promoting items that you want to move faster.

Promotion

To the consumer, promotion is the most familiar part of the marketing mix. This includes all of your marketing activities that let your customers know about the products you offer. Retailers typically advertise in newspapers and on radio and TV and might also have a social marketing campaign, marketing e-mails, or flyers. Promotional campaigns are directed to your targeted customer base.

Lean Marketing

When you are considering the marketing mix, many of the opportunities for Lean will be found in other areas that we will be discussing both later in this chapter and in the following chapters on merchandising and operations. However, in the "promotion" component of the marketing mix, I think it is useful to look at ways to make that more efficient, and in today's world, which is changing so rapidly in this regard, it is helpful to look at recent trends.

Trends

According to a presentation by Rohit Bhargava of the advertiser Ogilvy given at a 2012 Shop.org Merchandising Summit in 2012 [Bhargava, 2012], there are 12 transforming trends in retail, some of which can be considered applicable to Lean thinking:

- *Pervasive shopping*—Mobile devices have created new forms of retail integration, so we can shop anytime, anywhere. We need to focus on adding value to the customers by allowing them to purchase through multiple channels and integrating with third-party e-commerce networks.
- *Retail curation*—Consumers now have almost unlimited choice. The idea of curation is to narrow that choice and offer more personalized experiences to buy and share. Consumers now expect this as a "given" in retail, especially online. The explosion of stock keeping units (SKUs) in retail has made it much harder to manage more "piles" of inventory at brick-and-mortar retailers, leading to frequent stock-outs and potentially higher operating costs for the retailer. By potentially using concepts like postponement (final assembly only when the product is ordered or bought), technology, and e-commerce, we can better manage this challenge.
- *Being useful sells*—Today's consumers buy from companies that provide value by being useful. We now have "digital tactics" to educate consumers instead of having a more expensive "dialogue."
- *Analytics everywhere*—The concept of data analytics (the science of examining raw data to help draw conclusions about information) is upon us, with better

technology now available. The retailer can utilize these data to become more efficient in terms of its marketing mix, and consumers will use these tools, products, and online services to measure, track, and chart their lives, helping them to decide what products to buy.

A report by PSFK [Weiner, 2013] outlines current trends in retail, describing opportunities for companies to focus more on what adds value to the shopper. They state that today's shopper is more sophisticated, is willing and able to trade data for a better shopping experience, wants instant access to expertise, and takes advantage of technology.

So in effect, retailers should use their shoppers as "brand champions" instead of pure customers, and should let them sell to other shoppers on the retailer's behalf through today's social platforms. For example, a large UK-based grocery and general merchandise retailer offers customers loyalty points for sharing its products on Facebook. Upscale retailer Neiman Marcus has an app that provides sales staff with customers' preferences.

Education of the customer can be a key component of the shopping experience both before and after a purchase is made. An example of this is Motomethod, a motorcycle repair shop that invites customers to work on bikes themselves.

Some forward-thinking retailers even let their customers design or customize products to their liking. For example, Evolvex sells customizable modular furniture online at affordable prices, and Levi's offers "Curve ID," where customers can look online for the company's "bold curve jeans" that are designed to fit different types of figures based upon the customer's input (i.e., waist, seat, and fit requirements).

The Promotion Mix

When considering Lean in sales and marketing, we need to carefully consider the promotion mix and where it adds value. The promotion mix is made up of the following components (see Fig. 7.4):

- *Advertising*—This appeals to a mass audience at a reasonable cost and creates awareness and favorable attitudes.
- *Publicity*—This reaches a mass audience with an independently reported message.
- *Personal selling*—This involves dealing with individual consumers to resolve questions and to close sales.
- *Sales promotion*—This stimulates short-term sales and increases impulse purchases.

Advertising

When we advertise, two key steps are to determine the message content and decide on the specific media placement of the ad, and there is plenty of potential waste and clutter in both steps.

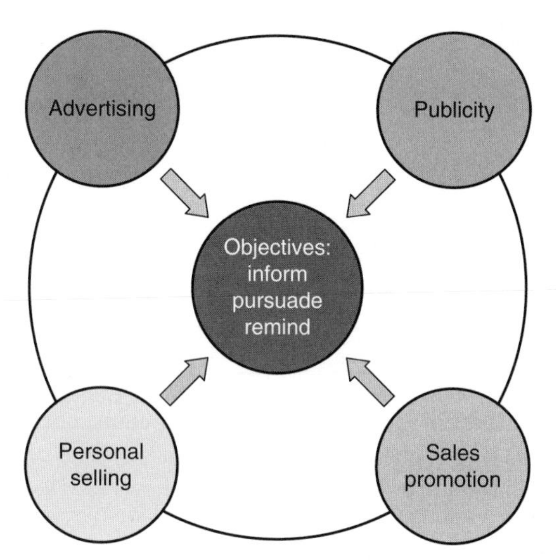

Figure 7.4 Promotion mix.

There is significant waste in advertising in the form of the portion of an audience that is not part of a company's target market, since advertising is appealing to a mass audience, and as a result, we should make sure that we've done a good job identifying our target audience and addressing its needs and desires in the ads.

Advertising also suffers from "clutter," which involves the number of ads that are contained in a single individual medium. Clutter is lower when a medium using a limited number of ads rather than many ads, as the latter significantly increases the amount of clutter.

Publicity

Publicity also needs to be clear, timely, and concise. For example, when you are offering new products, advertising without publicity will not get a good response, compared to advertising that is coordinated with a preintroduction news release, product samples, and testimonials.

Personal Selling

Personal selling refers to "one-to-one" selling. This is typically a very high-cost method for the retailer, but it offers great flexibility and control over content and placement. The ultimate goal of this type of selling is to deal with individual consumers, to resolve questions, and to close sales.

As a result, this part of the promotion mix can be laced with wasteful elements such as lack of qualified personnel, inadequate training, and limited access to and space for the information or items needed to answer questions and demonstrate products. We will cover this in greater detail later in the book when we discuss Lean store operations.

Sales Promotion

There are many types of sales promotion, including coupons, refunds or rebates, samples, contests, bonus or multipacks, and point-of-purchase displays. The advantages of promotions are numerous, such as to attract customer traffic and make quick sales. The types that provide value to the customers and are retained by them, such as calendars, pens, and the like, can help to create loyalty.

On the other hand, if a retailer is continually running promotions, the consumer can view that as reflecting a decline in product quality and believe that the store can't sell the items without promotions. This then becomes a waste of resources if it is not monitored carefully.

Coupons and other special deals, if used too often, can cause customers to stock up on items, resulting in the aforementioned inefficiencies of the "bullwhip effect" on upstream supply chain inventories. So this needs to be monitored as well and is a reason why Walmart and other retailers have adopted an "everyday low pricing" model.

On occasion, sales promotions can take customers' focus away from the product, resulting in negative effects on a brand's image and ultimately sales because the product's advantage hasn't really been developed with the customer.

Example: Waste in Sales Promotion A case in an Accenture Consulting white paper entitled "Solving Retail Problems Using Lean Six Sigma" found the following:

> A large US retailer determined the advertising portion of the weekly insert took 14 work days to create. The defect rate with its current process exceeded 90% resulting in extensive rework. Although the retailer never missed an insert, the process to meet printer deadlines was costly.
>
> After determining that more than 30 hand-offs were involved in the process of creating the insert, the Accenture team determined that the accelerated improvement approach was the most appropriate to help resolve the issues. Three major planning activities were necessary: determining the necessary team members, socializing the process and creating the agenda for the event itself. During the 12-hour session, 40 cross functional team members determined the as-is value stream map, the root causes of the process delays, the future state, and the steps and ownership needed to implement the improvement changes.
>
> The team implemented multiple solutions with the most important being a distinct timeline for each step in creating the insert, based on hard deadlines. The solutions were piloted in one class of product, and then implemented throughout the organization over the next few months. Once completed, the cycle time from the first touch by advertising was reduced to four days, and the defect rate decreased to less than 5 percent." [Curtis et al., 2008]

Trends in the Promotion Mix

When discussing the promotion mix, we can't leave out the current trends of using mobile commerce and social networking as cost-effective (and Lean) marketing tools with great reach. These are used by a variety of businesses, including e-tailers and "brick-and-mortar" retailers.

Mobile Commerce (M-commerce) E-mail has always been one of the most effective customer retention tools. As a result, retailers are now investing in directing e-mail toward the mobile consumer, since 44 percent of retailers' marketing e-mails are now opened on a mobile device.

Retail marketers are also focusing their mobile efforts on search engine marketing, as more than 20 percent of retail traffic now comes through mobile devices.

There may also be long-term potential for retail marketers in the use of Quick Response (QR) codes, two-dimensional bar codes that link a physical object to the mobile web via a smartphone scan, as well as location-based mobile advertising (e.g., a personalized message sent to a shopper while he or she is in a store) [Siwicki, 2013].

Social Networking Customers are also using social media to stay informed and connected to brands, offers, sales, and the like. Social networks are used to stay in contact with brands and retailers throughout the day. Therefore, as the use of social media sites continues to grow, so does the importance of a social media strategy for retailers.

Increasingly, customers are accessing their social networks with smartphones. According to Monetate.com, 80 percent of smartphone users access social networks on their devices, 55 percent of those users visit social networks on their devices once per day, and 96 percent of smartphone users have researched a product or service on their phone. As a result, the need for retail brands to be present on these social outlets is critical.

Social media strategies to promote your company and its products can include sharing new products that have just arrived, promotion of exclusive deals and offers, and sharing of employee recommendations.

To be truly cost-effective with social networking as part of your promotional mix, you need to be able to measure its success. These metrics should include:

- *Fan/follower growth*—The number of new fans gained during a certain time frame.
- *Engagement rate*—The number of user interactions (likes, comments, retweets, replies, and shares).
- *Response rate*—The percentage of user posts or questions that the administrator responds to.
- *Response time*—The average amount of time that the administrator takes to respond to user posts or questions.

- *User activity*—The hours and days of the week when users are engaging most frequently.
- *Shareability*—The number of shares and retweets that a post gets.
- *Interactions*—The number of interactions that a page or post receives and the types of interactions (likes, comments, retweets, replies, and shares) [Sassi, 2013].

Sales and Marketing Waste—Final Thoughts

In general, sales and marketing activities in any type of business tend to include wasteful activities. Ultimately, the customer suffers as a result. These wasteful activities may include the following:

- The mission of the company is not communicated or understood throughout the organization.
- Not enough market research is performed, and, if it is performed, it's not focused, defined, and/or prioritized properly.
- The marketing mix is incorrect for the target customers or, at the least, out of balance.
- Advertising and "brand awareness" campaigns don't really create much of a customer response.
- Media ads are reviewed multiple times by many executives, increasing costs and slowing down responsiveness.
- Valuable resources are spent on marketing communications, such as articles in publications and newsletters, that aren't read by the target audience.
- Results of marketing campaigns are not adequately measured and analyzed.
- Salespeople are spending too much time on the wrong customers or products.
- Information from service departments isn't properly communicated to management to help it improve the firm's offerings.

Lean thinking should always make us ask, "What value does your process create for the customer?," keeping in mind that both prospects and customers pay for products with their attention and time before they actually buy an item. So when your ad and product promotion campaigns don't generate a response, you're not giving the customers what they want.

Everything you do to find, win, and keep customers needs to create value in their minds, and that value needs to be measured in terms of the actions the customers take. If our sales and marketing processes are valuable to our customers (aside from our products and services), our marketing communications and newsletters should address the needs of prospects and customers, and should contain examples and cases that apply to the target audience because prospective customers want to get a feel for what your current customers are experiencing.

Location Strategy

As we've always been told, the three most important things in retailing are "location, location, location." This is largely due to the fact that the location decision is one of the most costly decisions you can make in retail. The decision is very complex, can cost a lot of money, and is a long-term decision with little flexibility once it is made.

Some of the criteria to be considered during this decision are:

- The size and characteristics of the local population, including demographic (quantifiable statistics for a given population) and psychographic (the study of personality, values, attitudes, interests, and lifestyles makeup) information
- The level of competition
- Access to transportation (buses, highways, and so on)
- Availability of adequate parking
- Attributes of stores nearby
- Property costs
- Length of lease agreement (if applicable)
- General and local population trends
- Legal restrictions that might apply, such as zoning and building codes, sign restrictions, and licensing requirements (e.g., a license to sell alcoholic beverages is required)

Retail Location Theories

There are quite a few general location theories that retailers use to help them determine the best location for their business. These include the following:

Checklist method—This method evaluates the relative value of a site in a systematic way by comparing it with other potential sites in the area. The checklist can be made up of factors such as traffic flow, population density, and income levels, which are given importance factor weights. The different options are then rated against the importance factors.

Retail gravity theory—This theory suggests that there are underlying consistencies in shopping behavior that can be analyzed mathematically to predict a location based upon the idea of the center of gravity of some retail shopping characteristics. An example would be what is known as Huff's Gravity Model, which is based upon the idea that the probability that a given customer will shop in a specific store becomes larger as the size of the store or the center grows or its distance from the customer shrinks.

Retail saturation theory—This theory looks at how the demand for goods and services at a potential selling location is being served by existing retail establishments when compared to other potential markets. Frequently an index of retail saturation is developed, that is, the ratio of demand for a product (households in an area multiplied by annual retail expenditures in an area) divided by the available supply (square footage of available retail facilities in an area).

Retail buying power index—This index is an indicator of a market's overall retail potential and is composed of the weighted measures of effective buying income, retail sales, and potential income.

Retail Location Lean Opportunities

As the location decision is truly strategic in nature, and essentially has the goal of maximizing revenues, any mistakes or miscalculations made can negatively affect the company for a long, long time. While a lot of this decision is based upon an "educated guess," doing the proper planning up front can save you big time down the road.

A Pitney-Bowes white paper on location intelligence [Pitney-Bowes, 2013] found that there are 10 common mistakes when it comes to retail site selection, all of which can lead to future waste. They are:

1. *Failure to understand the limitations of a forecasting model*—Some of the methods described earlier are very complex in a constantly changing retail environment. There are some factors that aren't easily measurable, like operations. Some companies can become too reliant on this type of tool. As any decision like this is a mix of art and science, you need to be careful to include intangible information, such as input from store personnel and management.
2. *Misunderstanding the dynamics of multiple sales drivers*—As they say, "All forecasts are wrong," so you need to be careful in considering drivers for your specific location and not try to "broad-brush" everything. You need to get a good feel for what people are shopping in a specific area and not oversimplify a target customer, as there may be multiple customer profiles for one location (e.g., a food store with a pharmacy).
3. *Unbalanced approach toward "optimal customers" and "overall populations"*—Make sure to focus on "quality, not quantity" by looking more at how many people in the area potentially fit your target customer profile rather than concentrating more on population mass (although that can work, too).
4. *Growing chains rendezvous with reality*—Some retailers, like many businesses, tend to overexpand too soon. Focus on things like productivity per square foot first. Recently, some retailers have begun to shrink their store footprint, as the best option for growth may come from in-store productivity increases (e.g., Home Depot is experimenting with stores that are less than half the usual size).
5. *Poor decision structure; enabling emotion over analysis*—This decision needs to be objective, and in many cases a group or a person may have too much invested to make an unbiased decision. Even though some variables may change during the decision process, sometimes it's like trying to stop a freight train. You need to balance this emotional investment with empirical analysis.

6. *Failure to recognize and capitalize on multichannel interaction*—The growth of e-commerce has made this decision more complicated, as you need to consider the relationship between the brick-and-mortar store and e-tailing. You need to consider the relationship of the physical store with order pickup, warehousing, and product showrooms. There is a trend for customers to order online and use the brick-and-mortar location more as a showroom.
7. *Competitive battle planning: stand your ground . . . or maybe not*—A concept known as *clustering*, in which similar retail businesses locate in the same area (e.g., McDonald's, Wendy's, and Burger King within a block of each other), can tend to draw you into an area based upon your confidence in your business. This needs to be looked at carefully and impartially to adequately quantify competitive impacts.
8. *Infatuation with form over substance in model selection*—A model may look and sound great, but that doesn't necessarily guarantee its success. You need to look at the model very closely, making sure that you validate it against reality, rigorously test it, and perform as much "what if" analysis as possible.
9. *The alluring market with elusive returns*—There may be some markets that are irresistible, like Manhattan; that appear to be a great opportunity that doesn't actually exist; or that have a much higher cost structure because of things like rent and logistics costs.
10. *Silos forever; not leveraging the real estate model throughout the organization*—A lot of the information and analysis that were gathered for the location decision can actually be shared with other areas and aid them in their decision making. The marketing and CRM (to be discussed later in this chapter) departments may acquire the same data and analytics and therefore duplicate the expense (or at least part of it) to the company. Information such as customer profiles, sales distribution, and market shares can be shared, and in fact some of the same analytical elements can drive both the location decision and marketing.

So while the location decision is strategic in nature, it's not hard to see how mistakes, miscalculations, and errors in this decision can lead to massive waste in the long term.

Human Resources Strategy

Despite the increased use of technology in retail, it is still a labor-intensive business, with employees being critical to the delivery of products and services and often making the difference between the success or failure of a business strategy.

Labor Productivity

Labor is a large percentage of retail expenses in terms of operations and selling and can contribute greatly to a company's efficiencies and competitive advantage.

One way to measure labor effectiveness is productivity. In retail, this would involve dividing the "output" (sales or profits) by the "inputs" (the number of employees). When properly implemented, Lean focuses on reducing cycle time and increasing throughput, which in many cases can be accomplished by increasing employee productivity.

Other aspects of labor attributes that can contribute to a retailer's success are satisfaction, morale, and commitment.

Instead of laying off employees during hard times, it's more effective to focus on a long-term vision of increasing employee productivity, satisfaction, morale, and commitment. Lean is one way to do this, as it's not about laying off employees, but about continuously training and empowering them and focusing their activities on value-added activities from the viewpoint of the customer.

Retail Productivity Case

You don't have to be a large retailer to implement Lean. In a case by UK-based consultant Alturos [www.alturos.co.uk], a small boutique coffee shop was struggling to become profitable so that it could expand to other outlets. The consultants worked with the client's staff to review all activities and found that:

- The staff members often had to ask more experienced or senior staff members for instructions and information, slowing down service to customers.
- The kitchen area layout appeared to be efficient, but after some analysis, the consultants found that the layout in some areas had excessive congestion, resulting in slower service.
- The shops frequently ran out of decaffeinated coffee. The assumption had been that this was due to the supplier, but the team collected data showing that demand had steadily increased.

The team recommended a staff training program to eliminate the constant questions from one staff member to another, designed a new layout to remove the congestion, and designed a real-time replenishment system, not just for decaffeinated coffee, but for *all* fast-moving lines to prevent stockouts and improve sales. Perhaps most important, the owner gave the team his full support.

The team members were motivated and engaged in the whole process as it was implemented over the following months, especially because they had been involved in the improvement process themselves. As a result, the shop's sales and profits were increased significantly.

This simple, yet effective coffee shop example shows how Lean can be used to improve labor productivity in retail.

Lean Failure

Over the many years that I have been helping companies become Lean, I have read (and heard) that well over 50 percent of Lean initiatives (at least in the

United States) have failed. From my experience, I tend to think that this is accurate. I believe that the primary reason for this is a lack of the proper culture to support the major—and, in some cases, radical—changes required.

Lean initiatives fail for a variety of reasons. In many cases, management is not willing to give up some control to workers, dedicate resource time, or spend money for training and improvements. In other cases, Lean is just looked at as a "fad" that will go away or only a short-term program.

Most U.S. companies seem hesitant to spend the time and money on the training required to become Leaner, and they lag far behind other countries in the amount of training, in general, that they provide for employees. While there are a few bright spots, such as General Electric, with its Croton-on-Hudson John F. Welch Leadership Development Center facility, most companies offer very little in the way of training and support. In general, the average annual training hours per employee in the United States is much lower than in most other developed countries.

Implementing Lean

Lean also requires both a top-down management commitment to change and a bottom-up groundswell of participation and ideas. Without them, it is a losing battle. The culture has to encourage and create a team-based continuous improvement mentality.

In the United States, the more common way to initiate change has been to have consultants come in who "borrow your watch to tell you what time it is," then leave reams of reports, slide shows, and the like for the client to review and interpret on its own. A myriad of problems can result from this methodology, ranging from no direction on how to implement the improvements to a general lack of enthusiasm and support from the people who are being asked to change based on suggestions from an outsider.

A more effective way to implement Lean, I believe, is through a train-do method. In this way, the trainer or consultant is more like a facilitator who teaches the employees the basic concepts and tools, but lets them create and direct the activities (with a little management oversight, of course). That way, after the trainer or consultant is gone (which is always the case eventually), the workers can continue on the Lean journey.

Middle and upper management still does much of the higher-level "out-of-the-box" thinking using the tools described in this book, such as value stream mapping (VSM), which is used to analyze and design the flow of materials and information in order to help identify opportunities; however, input and actual implementation has to include everyone in the organization.

In many cases, there also seems to be a gap between training in general Lean concepts and teaching people how to actually begin the improvement process (and to convince management that it can benefit the company's bottom line, not just help employee participation and morale). That is why doing a Lean opportunity assessment (LOA), discussed later in this book (and in the template included in Appendix B), is a great place to start [Myerson, 2012].

Working Together

Levy and Weitz point out in their book *Retailing Management* [Levy and Weitz, 2012] that there is a "human triad" of sorts in retail, with HR professionals, store managers, and employees needing to work together to reach their full potential (see Fig. 7.5). They mention that some retailers, such as Wegmans and Home Depot, have been leaders in this area.

For example, Wegmans grocery stores' HR practices and its overall stakeholder orientation state that the treatment of employees as strategic assets constitutes an effective approach to social responsibility and good business. They have five values:

1. They care about people.
2. High standards are a way of life, and they pursue excellence in everything that they do.
3. They make a difference in every community that they serve.
4. They respect their people.
5. They empower their people to make decisions that improve their work and benefit their customers and their company.

Hiring and Developing Employees

An aspect of the HR strategy that is of particular importance for Lean thinking is the selection and development of the employees.

During the hiring process, some retailers will actually allow team members to interview candidates and give their opinions as to whether the candidate is a good fit for the existing team.

According to the Whole Foods grocery chain Web site, "Team member participation in group interviews is one way we put our culture of empowerment into

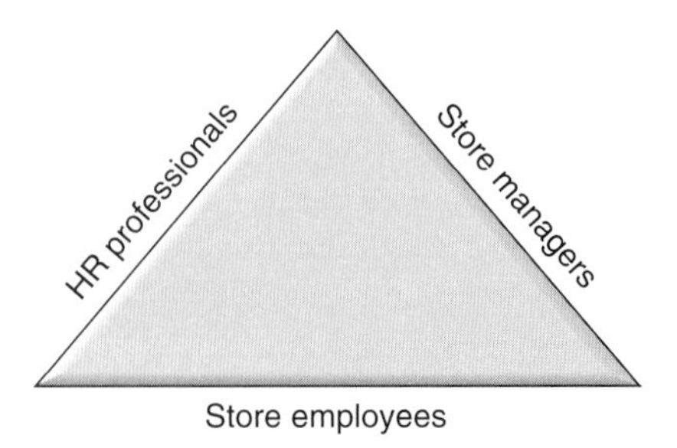

Figure 7.5 Human triad.

action. The team interview process is an effective way to make hiring decisions because the diversity of participants brings all of the different aspects of the roles and responsibilities of the position to the table. In addition, the process educates both the interviewers and the candidates by giving insight into all the expectations and challenges of the job [Whole Foods, 2013].

Hard Rock Cafe has lower employee turnover than the industry overall because of its:

- Thorough screening at hiring
- Tolerant, even accepting culture of diversity
- Great benefits package
- Effective indoctrination, training, and empowerment, including an initial two-day training class
- Focus on teamwork and even outside volunteer work [Heizer and Render, 2011]

Whole Foods and Hard Rock Cafe are two very successful businesses in their industries, which I think shows that tangible results can be obtained by being meticulous in terms of the hiring and training of employees.

Motivating Employees

As they say, "A happy employee is a productive employee." Compensation-based incentives can be an important motivator of employees, especially in conjunction with a continuous improvement program of any kind.

There are a variety of ways to deliver this particular kind of benefit. They include:

- *Commission*—A fee or percentage given to a sales representative or an agent for services rendered.
- *Profit sharing*—Distribution of profits to employees.
- *Bonus*—Cash or stock options (individual or group).
- *Incentive plans*—Typically based on production rates in manufacturing.
- *Knowledge-based systems*—Rewards for knowledge or skills attained.
- *Gainsharing*—A management system used to increase profitability by motivating employees to improve their performance through involvement and participation. As performance and profitability improve, employees share financially in the improvement (i.e., gain). Gainsharing's goal is to improve performance and eliminate waste (time, energy, and materials) by motivating employees to work smarter, as a team, rather than just working harder. This has historically been used in manufacturing, but it is a great tool to use in Lean retail.

By motivating employees to be more productive with rewards such as those that we've just discussed, you're also encouraging them to work Lean. When employees are as passionate about productivity improvements as their managers are, they'll come up with many ways to eliminate waste everywhere they see it in their daily activities.

Empowerment

Empowerment typically refers to the process of allowing employees to have input into and control over their work, as well as the ability to share suggestions and ideas about their work and the organization as a whole. An empowered employee tends to be more committed and loyal.

A self-directed work team is a group of empowered individuals working together to reach a common goal. Self-directed teams may be organized for long-term or short-term objectives and are especially useful in a Lean program. They are effective because they:

- Provide employee empowerment.
- Ensure core job characteristics.
- Meet individual psychological needs.

Again referring to the Whole Foods Web site, we find that in its case,

> The process of selecting a new Whole Foods Market team member doesn't end with the hiring process! During your orientation period (typically 30–90 days) you will attend orientation and training classes, and you and your team leader will discuss your progress.
>
> If your team leader then recommends that you be considered for team membership, your fellow team members will vote on whether to add you to their team. The team vote process empowers team members to share in the building of a quality team and strengthens communication. The criteria used to vote a new member onto a team are positive job performance, adherence to policies and procedures, excellent customer service skills, and teamwork. [Whole Foods, 2013]

To maximize the effectiveness of a self-directed work team, managers should make sure that those who have legitimate contributions to make are on the team, provide adequate support, ensure that the team members have the necessary training, endorse clear objectives and goals, and establish financial and nonfinancial rewards. Perhaps hardest of all, supervisors must release control. (There still should be some kind of reporting structure, such as a "steering committee" of executives.)

Flexibility and quality are critical to Lean retail, and processes that empower employees to customize the services provided to meet or exceed customer expectations are very powerful. Traditional customer service models usually have a set structure in which employees lack the freedom to think on their own. For example, there may be a list of tasks and paperwork to complete in response to a complaint. The cost of such inflexibility is an unhappy customer.

When using Lean and Lean Six Sigma in retail, employees can focus on and measure the levels and conditions of service delivered to individual customers instead of following a strict set of procedures. This type of freedom creates an environment in which creative solutions and service quality flourish.

As a result, customers will receive the information and service they desire within a reasonable amount of time. In many cases, quality of service is improved so much that it may result in repeat business that otherwise might not have occurred.

Information Systems and Supply Chain Strategy

An efficient supply chain can benefit a retailer by giving it a strategic advantage over its competitors, improved product availability, and greater return on investment. Walmart has been a great example of this over the past 20 years by investing heavily in systems and continuously improving processes, both internally and externally, and by collaborating extensively with suppliers.

A truly Lean supply chain enables a retailer to integrate its various suppliers, manufacturers, warehouses, stores, and transportation intermediaries into one "value" chain. This will allow the retailer and the wholesaler to have "the right product at the right place at the right time."

As pointed out in the earlier discussion of the value chain, the results are lower systemwide costs and greater customer service, with fewer stockouts and greater assortments than the firm's competitors. Through efficient management of the supply chain, there will also be the opportunity to reduce inventory and transportation costs.

The key to enabling and controlling a lean supply chain process in retail and wholesale is information. Let's explore some of the technology that can make this happen.

Retail Technology Basics

Cash registers have been around since the 1870s, but it wasn't until the 1970s that computer-driven cash registers were introduced.

> The first computer-driven cash registers were basically a mainframe computer packaged as a store controller that could control certain registers. These point of sale systems were the first to commercially utilize client-server technology, peer-to-peer communications, Local Area Network (LAN) backups, and remote initialization.
>
> In the late 1980s, retail software based on PC technology began to make its way into mainstream retail businesses.
>
> Today, retail point of sale systems are light years ahead of where they began. Today's POS systems are faster, more secure, and more reliable than their predecessors, and allow retailers to operate every facet of their business with a single, integrated point of sale system. [RetailSystems.com, 2013]

In retail today, more of the technology is being implemented in the supply chain rather than in the stores, offering great increases in productivity. Of course we have bar code scanning at checkout (and self-checkout), credit card payment, and other such systems in the stores, but most of the technology is back-room-oriented to help firms better manage inventory in the supply chain. This is in contrast to the world of e-commerce, where the buyer is more informed and has access to more options via the Internet (e-tail storefront, research, m-commerce, and so on).

As there are more applications in the supply chain, let's explore that area first.

Retail Supply Chain Technology (Information Flow)

Over the past 30 years, retailers have increased the use of technology to support and improve their supply chain management processes. The improved flow of information touches all areas of a retailer's business and flows both ways to help in the planning and delivery of merchandise (see Fig. 7.6).

This started in the 1980s with electronic data interchange (EDI) and the use of bar code scanning to keep more accurate track of sales and inventory, and it has helped to increase the speed of retail supply chain processes. This technology has helped retailers to have better information accuracy and reliability, shorter lead times, lower stockouts, faster response, and lower costs.

As pointed out in "Use of Information Technologies in Retail Supply Chain: Opportunities and Challenges" [Ying Xie, 2009], IT technology applications in the retail supply chain are increasing. The major technologies, as seen in Fig. 7.7, are the following.

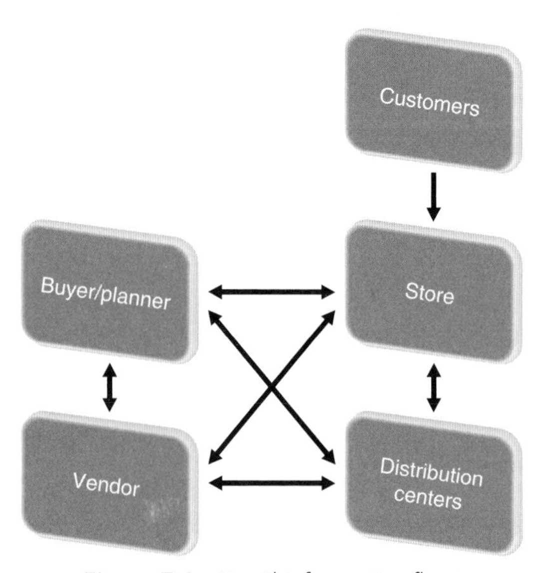

Figure 7.6 Retail information flow.

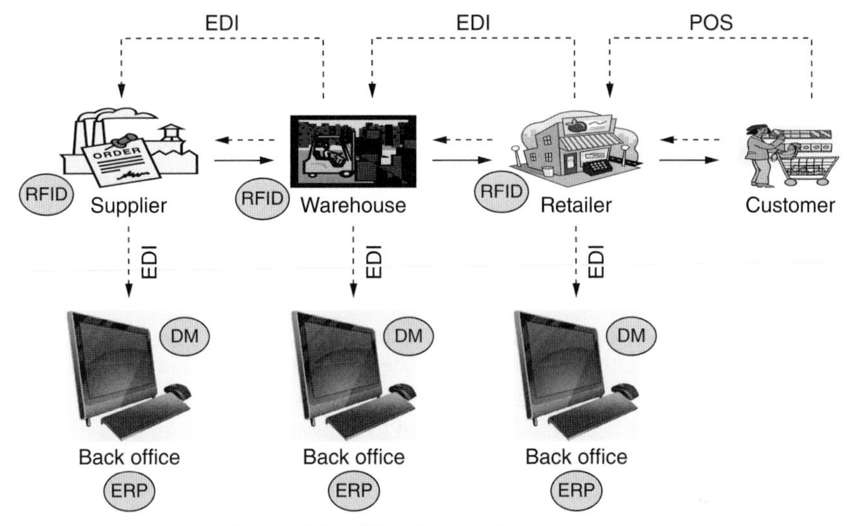

Figure 7.7 IT in the retail supply chain.

Electronic Data Interchange

EDI is a standard format for exchanging data between businesses. An EDI "envelope" contains a string of data elements, each of which represents an individual piece of data, such as a price, a product model number, or something similar, and is separated from the next element by a delimiter. The entire string is called a data segment. One or more data segments surrounded by a header and trailer form a transaction set, which is the EDI unit of transmission (i.e., a message). A transaction set often consists of what would usually be contained in a typical business document or form. The parties that exchange EDI transmissions are usually referred to as trading partners.

EDI is primarily used to place electronic purchase orders, generate bills of lading/freight bills and invoices, transmit sales and inventory data, and generate advanced shipping notices (ASN).

Traditionally, EDI moved through what are known as value-added networks, which work as a kind of mailbox where the receiving partner can consolidate customer data such as purchase orders and retrieve them multiple times per day. In 2003, Walmart announced that it would start doing Internet EDI with all its trading partners, which created major difficulties for the value-added networks. However, others decided not to make the switch at that time because of the costs involved.

In the future, the question is not if, but when there will be a significant migration to the Internet for EDI, at least for larger organizations.

Point of Sale

At the point of sale (POS), retailers use any of a range of possible methods to calculate the amount owed them for items, such as a manual system, weighing machines, scanners, or an electronic cash register. The retailer provides the hardware and options that the customer can use to make payment. The merchant will also normally issue a receipt for the transaction.

An automated POS system reads the bar code from items and enables dynamic update of item sales and inventory data as sales occur. POS systems are widely used for store-level transactions, credit and/or check-cashing authorization, credits, refunds and exchange issues, promotional inventory tracking, item and price lookups, sales tax calculation and reporting, and dynamic accounts receivable updating.

Reports in POS systems are most helpful in gaining better control of your business. The information in the reports can help you determine what products are selling best at what time, ways to arrange shelves and displays, and which promotions are working best and which need to be changed.

More recently, POS data are being used to integrate partners in the retail supply chain and used extensively for improving forecasting accuracy and reducing lead time (both leading to waste). In fact:

> More and more retailers are sharing sales data as POS transactions with more and more vendors. Mass merchants have led the way, and many now share POS data with their suppliers. SPS Commerce estimates as many as 120 retailers have implemented collaborative programs in which they share store-level sales data, inventory data, and margin information. Drug retailers and grocers have not been as forthcoming, but there is movement even in those sectors. Drug store chains CVS Caremark Corp. and Walgreen Co., for example, do, as do many of the grocers including The Kroger Co. and Safeway Inc. Whenever retailers and suppliers share POS data, the general outcome is a far more strategic relationship with advantages for both.

As was pointed out in Chap. 3 when describing various collaborative relationships for retailers,

> Manufacturers and suppliers could use the data for such things as demand planning, improving supply chains to avoid stock outs and to better understand and act on buying habits so they can more accurately allocate advertising dollars. [Bachelder, 2012]

The challenge now is for small- and medium-sized retailers and wholesalers to begin to share these data (and in more detail) with suppliers and partners to improve the efficiency and lower the cost of the entire supply chain for all participants.

It's certainly worth the effort, as POS systems can help retailers meet customer demand and boost operational efficiencies, and the retailers are also rewarded with increases in customer loyalty, same-store sales, and gross margins. POS systems for retail can enhance a firm's brand by better managing data on customers, products, POS, inventory, pricing, and store-level sales.

Radio-Frequency Identification

Radio-frequency identification, better known as RFID, is a system that tracks material using tags that respond to radio waves. RFID tags are integrated circuits that have a tiny antenna and are not easily noticeable.

RFID tags can be active or passive. An active tag has a battery and transmits its ID signal. A passive tag is cheaper and smaller because it has no battery and obtains its power via the transfer of energy from radio frequencies that are emitted by a "field reader" RFID (field readers are what you see in the entrance and exit of a retail store to look for merchandise theft). As a result, passive tags have to be within range of a reader.

The primary difference between RFID and bar code technology is the data exchange and scanning process. RFID can track the quantity of the items without a scanner and send the information to the receiver located in the retailer's warehouse, making the system more flexible and efficient.

Like Universal Product Code (UPC, or bar code) labels, RFID tags are often used to uniquely identify the object they are attached to, but unlike UPCs, RFID tags don't need to be scanned with a laser scanner. Instead, they can be identified and recorded by placing the tag within the range of an RFID radio transmitter, as noted earlier. As a result, it is possible to scan several items quickly or to locate a specific product that is surrounded by many other items.

RFID has been used primarily in warehouses for data capture, but of late, some retailers are using it to manage in-store inventories. Marks & Spencer, a large UK-based retailer, recently announced that it is rolling out RFID for all the apparel and hard goods at all of its stores. This is as a result of a successful pilot at 120 stores in 2006, where:

> Employees at those RFID-enabled stores have been equipped with handheld RFID readers to take stock of inventory in six specified departments where having a variety of clothing sizes on the floor is most important. In each store, it takes the six departments about eight hours to count inventory manually, says M&S spokesperson Olivia Burns. Using the RFID system, however, the same process—done every few days per store—takes about one hour.... The inventory-taking process has become faster and more efficient, and store employees now know when an item is running low in inventory or out of stock, necessitating a reorder. In the event of a shortage of a specific size, style or color on the floor ... the system alerts employees so they can reorder the item in question. [Swedberg, 2006]

Marks & Spencer said that it has informed customers about the RFID tags before they purchase the items, stating that they are using this system to improve customer service and efficiency. Information about the tag is on the label, and leaflets are also available in the store.

So since many larger retailers like Marks & Spencer and Walmart have found that they can reduce waste in their supply chain with RFID technology, the future use of RFID will expand to other stores as well. This might have an even broader application and payback as the technology keeps getting less costly, enabling small- to medium-sized retailers to implement RFID in their stores as well.

Data Mining and Data Warehousing

Data mining (DM) is the process of analyzing data from different perspectives and sources (internal and external to your business) and summarizing them into useful information that can be used to increase revenue and/or cut costs and is stored in what is known as a data warehouse. DM software is a tool for analyzing data that allows users to examine data from many different dimensions, categorize them, and summarize the relationships identified.

Data warehouses can be a source of information used for making strategic decisions like developing competitive strategies, identifying market opportunities, launching new products, and product positioning. They can also be used for tactical decisions such as sales forecasting; direct marketing; customer acquisition, retention, and extension purposes; and marketing campaign analysis.

Retail Enterprise Resource Planning Systems

Additionally, operations managers can use similar data for decisions such as supplier, logistics partner, and transportation selection; purchasing/buying; and inventory control.

Enterprise resource planning (ERP) systems, also known as retail merchandise management (RMM) systems and merchandise management systems (MMS), are multimodule, integrated software systems that help to manage and support a broad spectrum of day-to-day business activities for the retailer. While they cover basic functions such as finance, payroll, and the like, it is most critical that they cover some core areas of the business. These are:

- *Inventory management (also known as merchandise management)*—The basic functionality related to on-hand or in-transit inventory; it can also include supply chain execution "add-on" modules to an ERP system, such as:
  - *Warehouse management systems (WMS)*—Software that controls the movement and storage of materials within a warehouse, including the shipping, receiving, putaway, and picking processes.
  - *Transportation management systems (TMS)*—Software for planning and executing inbound and outbound transportation processes.
- *Inventory optimization (also known as merchandise or assortment planning)*—Tools used to assist in the buying and selling decisions that affect inventory. These tools typically consist of add-on modules to an ERP system called a supply chain planning system that helps organizations to more accurately forecast demand, plan procurement and production, balance inventory across the supply chain, and maximize revenue.
- *Revenue management*—Tools used to make sure that inventory is sold at the right price by looking at history and trends. This function can also be used to

help with promotions and to determine markdowns on discontinued or out-of-fashion products.

- *Sales management*—This is part of the front-office system that is used at the cash register and typically includes POS systems (either theirs or a partner's), which were discussed earlier.
- *Reports and inquiries*—Daily reports to help upper management make good decisions [Sika Retail, 2006].

The use of any or all of these types of systems can enable a better flow of information and communications and, as a result, create a Leaner, more efficient supply chain.

Retail Supply Chain and Logistics Management (Material Flow)

There are many definitions and understandings of the supply chain. Technically, supply chain management includes the logistics function, which includes the transportation and distribution areas; hence I and many others often refer to it as "supply chain and logistics management." However, others take a very narrow definition that is primarily focused on procurement and purchasing.

It is more effective, especially from a Lean perspective, to take a very broad view similar to the definition from the Council of Supply Chain Management Professionals (CSCMP): "Supply chain management encompasses the planning and management of all activities involved in sourcing, procurement, conversion, and logistics management. It also includes the crucial components of coordination and collaboration with channel partners, which can be suppliers, intermediaries, third-party service providers, and customers." [Council of Supply Chain Management Professionals, 2013]

In retail, the supply chain comprises all the parties that participate in the process, which include manufacturers, wholesalers, third-party service providers (that is, shippers, order-fulfillment houses, and the like), and retailers (see Fig. 7.8).

As discussed in Chap. 4, many retailers are seeking closer relationships with suppliers using collaborative programs like Collaborative Planning, Forecasting and Replenishment (CPFR), Efficient Consumer Response (ECR), and QR, which have resulted, in many cases, in Leaner, more responsive supply chains.

There has also been a growth in the use of third-party logistics providers (3PLs), where experienced supply chain professionals work closely with retail businesses to improve service, optimize distribution and transportation networks, and streamline their global supply chains. These logistics specialists work with retailers of all sizes to ship and warehouse merchandise.

The Internet is an increasing part of supplier–retailer communications. In fact, many manufacturers and retailers have set up dedicated sites exclusively to interact with their channel partners, where vendors can access sales data and inventory reports, accounts payable figures, invoices, and performance report cards.

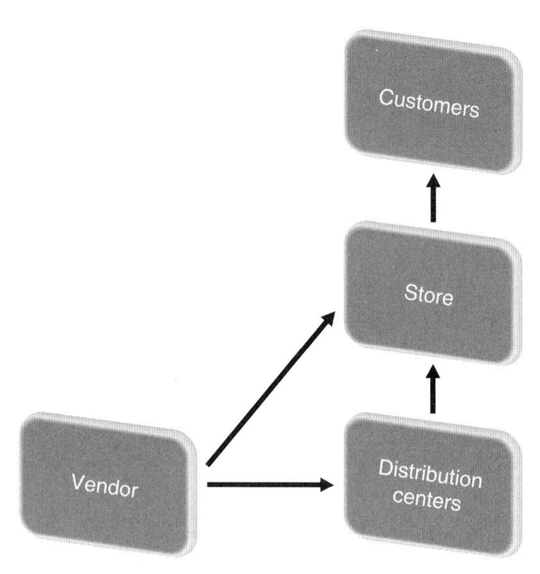

Figure 7.8 Retail merchandise flow.

Retail Lean Supply Chain Example

A great example of a retailer using its supply chain for a strategic advantage to eliminate waste is Zara, a worldwide fashion retailer. It employs the following strategies:

- *Quick response to demand*—Zara uses a demand pull model in its inventory and supply chain management. It creates more than 1,000 designs every month based on store sales and current trends and keeps a close eye on customer spending in the stores to evaluate and understand what types of designs are being purchased and then try to build that information into its next designs. It also receives timely information from store managers through handheld devices connected to the corporate office for additional feedback.
- *Small batch size short lead time production*—Zara has high inventory turns, as it produces small quantities of every product that is sourced locally at Zara's single centralized design and production center, which is attached to Inditex (Zara's parent company), headquartered in Spain. This gives Zara the opportunity to quickly understand what designs are successful. It is also a great way to explore new designs and understand their acceptance rate in the market. This also heavily reduces the risk of producing large quantities of something that the customer does not want and being left with inventory that the firm will have to discount heavily. Even though it might seem expensive for a retailer to invest in different designs, Zara optimizes by using the same material, only in different ways.
- *Central distribution center*—Zara has very strong IT systems that support its distribution. All clothes are shipped from the central location in Spain, then

distributed to different countries and stores based on the individual requirements and needs of the particular retail locations [Ferdows, 2004].

Transportation and Warehousing

The logistics function, primarily through transportation and distribution, can add a great deal of value to the retailer and its customers by efficiently:

- Managing inbound transportation from suppliers
- Receiving and checking merchandise
- Storing or cross-docking merchandise
- Getting merchandise floor-ready, including ticketing and marking
- Putting items on hangers
- Preparing to ship merchandise to individual stores
- Managing the outbound transportation (delivery) function

However, if Lean continuous improvement methodologies are applied, this process can be used strategically by the retailer to:

- Have more accurate sales forecasts by combining forecasts for many stores serviced by one distribution center.
- Carry less merchandise in the store.
- Enable quick replenishment to avoid running out-of-stock.
- Save on operating costs, as retail store space is more expensive than space at the distribution center.

We will discuss Lean in both the store and the distribution center in more detail in a later chapter, but in general, as Peter Bradley says in "The Skinny on Lean," it's a good idea to follow a five-step process as a guide to implementing Lean principles, which can be applied to the distribution environment as well as to manufacturing:

- *Identify what your customers expect and determine what value you add to the process*—For distribution and logistics, that usually means greater velocity. What it doesn't mean is a lot of handling. Distribution people assume all the handling they do adds value, but customers don't see it that way. "No customer asks if a product has been touched a lot," Womack says. "Most people just want their product. All those touches from a customer standpoint are irrelevant. From an end customer standpoint, less logistics is better."
- *Plot the value stream*—Identify all the steps involved in moving goods through the system. Womack and Jones encourage the use of VSM—literally diagramming all the steps in the distribution process, from order to delivery. That diagram may help you spot activities that add no value so that you can eliminate them.
- *Make the process flow*—Dismantle any roadblocks that prevent the free flow of materials through the facility.

- *Pull from the customer*—The lean system is a pull system, drawing materials and merchandise into the distribution network based on what customers want (not on hazy forecasts). Many retailers use an allocation system, which will work if demand is steady. However, in many cases it's not and therefore it's wise to generate orders from the store based upon POS sales data.
- *Pursue perfection*—Root out any remaining waste. Then do it again, and again, and again [Bradley, 2006].

Outsourced Logistics Services

There are many aspects of the logistics function that can be outsourced, as in many cases it's not a core competency of a retailer, especially a smaller one. Not all retailers are as large as Sears, which has a dedicated logistics division.

Aspects of logistics that can be outsourced include transportation, warehousing, and freight forwarding, and there are also integrated 3PLs that can "do it all" for the retailer.

> While some firms view logistical outsourcing as a means to reduce cost, on a much broader basis it can also enhance resource utilization and manage risk. From the cost perspective, logistical outsourcing can reduce cost because most firms that offer logistics outsourcing arrangements are nonunionized. By outsourcing unionized manufacturers can shift material handling, warehousing, and transportation activities to a logistics service provider at a lower wage rate. Logistics outsourcing can also result in more efficient processes and improved resource utilization by assigning a supply chain activity to a partner that can take advantage of economies of scale, scope, or expertise. Specifically, a supply chain partner that primarily focuses on a supply chain activity such as warehousing, transportation, technical expertise, or process expertise can likely complete the task with less risk and a lower cost. Logistics outsourcing can also improve resource utilization by pooling capacity across a number of clients, resulting in a lower risk of not being able to meet customer demands due to a surge.... These benefits contribute to the outsourcing firm by reducing risks related to uncertainty, capacity, process, and expertise. [Bowersox et al., 2013]

Customer Relationship Management Strategy

In today's customer-driven economy, retailers must move from product-based campaign marketing to a customer-directed approach.

CRM Defined

CRM is an approach that can help organizations serve their customers better. CRM helps a firm to identify valuable customers, assess their needs, and provide more personalized service to build loyalty. It also streamlines the handling of inquiries and requests, resulting in greater operational efficiency and faster responses to customers.

Not all customers are equally profitable, and more and less profitable customers need to be treated differently.

Retailers are now concentrating on providing more value to their best customers, using targeted promotions and services to increase their share of wallet (the percentage of the customers' purchases that is made from the retailer).

There is also a class of software, also referred to as "CRM," that provides highly automated tools to generate and track sales leads, the performance of individual products or sales professionals, and the results of sales campaigns on a wide range of parameters.

CRM Goal

The goal of CRM in general is to make sure your organization is keeping customers satisfied, identifying and solving problems to maximize revenue and profits.

This has become especially challenging with the emergence of the Internet, which is another source of customer information. As a result, it is changing the way we manage our customers.

CRM Best Practices

According to a recent survey of more than 100 companies, best-in-class retailers are 60 percent more likely than their peers to have an enterprisewide CRM application. In addition, they are 21 percent more likely than the industry average and 73 percent more likely than laggards to have a loyalty mailing application, and are 41 percent more likely than the industry average and twice as likely as laggards to have a loyalty database hosting application. All of these enablers are connected through an enterprisewide CRM system that allows associates and managers across the organization to gain visibility into the customer database, and helps to promote loyalty campaigns at the store level by aggregating data from multiple sources [Aberdeen Group, 2010].

From a Lean perspective, building better customer relationships to increase sales and profits requires new business process approaches, new analytics, and new integration of systems (as well as new strategies and tactics) that can enable employees to capture and leverage customer information.

According to the same Aberdeen Group survey, loyalty initiatives in retail are a major driver of success. They tend to perform at a higher level in terms of many metrics of success, such as annual growth rate and customer retention. The metrics show that customer loyalty programs are a direct indicator of success, since they help firms to acquire new customers, retain current customers, and reactivate dormant customers.

The Aberdeen report also recommended that retailers adopt some key loyalty initiatives:

- Create a customer loyalty program road map.
- Conduct customer wallet share and market basket analysis.
- Apply a rules-based and POS-integrated customer loyalty system.
- Develop a multitier loyalty program for the most profitable customers.
- Measure the net profit margin impact of customer loyalty expenditures.
- Upgrade the loyalty tool infrastructure on an annual basis.
- Improve automated enrollment of customers at the POS.

The best-in-class performers in the survey grew their average customer spending and retention (16 percent versus 4 percent for the industry average) and decreased customer attrition (5 percent versus 1 percent for the industry average) annually.

CRM Process

So how exactly do retailers do this? They need to understand that CRM is an iterative process that turns customer data into customer loyalty through four basic activities (Fig. 7.9):

- *Collecting customer data*—These can include data from previous transactions, customer contacts (visits to the Web site or the store, inquiries to the call center, and so on), demographic and psychographic data, and customer preferences.

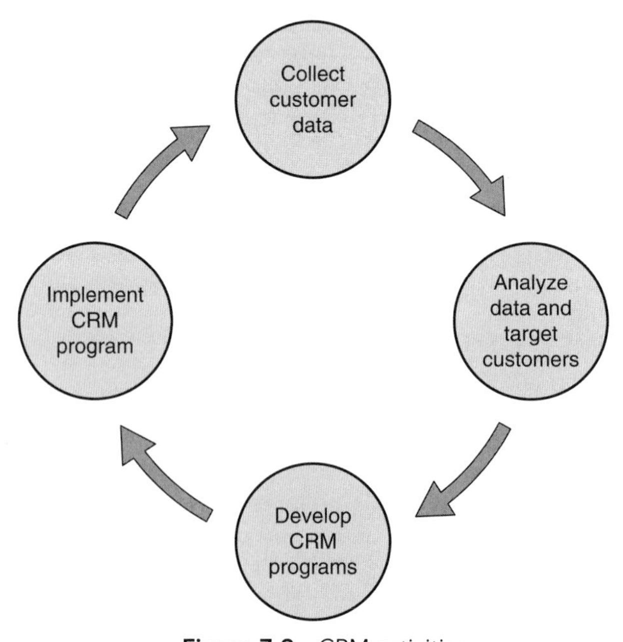

Figure 7.9 CRM activities.

The data can be gathered in-store by asking for identifying information and frequent shopper cards as well as online from Internet purchases.

- *Analyzing the customer data and identifying target customers*—The data gathered have to be analyzed and converted into usable information to develop customer loyalty programs. The DM tools mentioned previously are a great asset in this regard.

  Some of the specific tools that can be used are:

  - *Market basket analysis*—This focuses on what items are bought during a single shopping visit.
  - *Market segment identification*—This uses past behaviors to forecast future purchases, profitability, and the costs associated with serving groups of customers. A great way to do this is through what's known as *RFM analysis* (recency, frequency, and monetary value).

These types of tools allow the retailer to identify market segments and ultimately customers to target that are the most profitable.

- *Developing CRM programs*—The goal is to design a program to retain the best customers and to turn good customers into best customers (and possibly get rid of bad customers). To do this, retailers often develop frequent shopper programs (like Kohl's, which is very successful) and special customer services and personalization (such as Nordstrom's, which is legendary for sending thank you cards, making home deliveries, setting up personal appointments, and making phone calls alerting customers to upcoming sales).
- *Implementing CRM programs*—While CRM systems and management are important, there must be a focus on customers, requiring a fundamental shift from being product-centric to being customer-centric.

CRM Status

In general, retail has been slow to adopt CRM practices.

> Most of it has been focused on store-based efforts like branded credit cards and frequent-shopper programs, some of which have taken root and been extended year after year.
>
> At the high end of the sector, several creative new CRM strategies have emerged. Both Nordstrom and Prada, for example, are experimenting with "clienteling," which means automating the "black book" traditionally maintained by store sales personnel on their own customers, containing purchase history, preferences and contact information. Sales people will be able to pull up customer profiles and email their best customers about upcoming events like trunk shows or the arrival of new merchandise.
>
> At the mid-level of the retail sector, the focus is on loyalty programs. In the grocery business, early efforts produced discount cards, that offered little more than the equivalent of manufacturers' coupons for periodic specials.

> The Food Emporium chain, a division of A&P, launched in March 2003 a new program called Gold Points. The points can be redeemed for cash savings at purchase or for other rewards like travel or consumer electronics. [Food Emporium has since replaced the program with other loyalty programs such as Rewards Online Mall and Fresh Club Card.]
>
> The next stage for retail CRM, after loyalty programs, is likely to be multi-stage marketing, where stores track buying patterns and develop up-sell and cross-sell strategies accordingly.
>
> As long as the retail sector is focused on cost savings, versus revenue enhancements, its customer strategies will be limited." [Stevens, 2003]

While it is important to focus on eliminating non-value-added activities, retailers also need to have a plan to increase value to their customers (and revenue) in order to attain the true benefits of Lean. CRM is one way to do this and, as mentioned, is still evolving in retail.

After strategy comes implementation, both tactically (merchandise management) and operationally (store and warehouse operations), which are the subjects of the next two chapters.

CHAPTER 8

Being Lean: Merchandise Management—Planning, Buying, Pricing, and Communications

Merchandising consists of the activities involved in acquiring goods or services at the appropriate time, place, price, and quantity to enable a retailer to reach its sales and profitability goals.

Merchandise "management," therefore, is the analysis, planning, acquisition, handling, and control of the merchandise investment of a retail operation.

Therefore, it's no surprise that merchandising decisions can have a huge impact on a retailer's performance.

Lean Merchandise Management

In general, the application of a Lean approach in retail allows a company to reduce costs, increase efficiency, shorten execution time, decrease waste, increase profitability, and lower inventories. It also contributes to greater customer satisfaction, improved product quality, and increased staff morale.

When applying Lean to merchandise management specifically, we need to think about the simplification and standardization of work, moving the supply chain process from a demand push to a customer pull form of replenishment, and removing bottlenecks that limit overall capacity and throughput in the supply chain.

We especially need to consider the identification and elimination of the eight wastes in terms of time, effort, materials, motion, and movement.

According to a McKinsey & Company survey, the effects of Lean thinking in retail are:

- Increased sales by 10 percent.
- Reduced labor costs by 10 to 20 percent.
- Reduced inventory by 10 to 30 percent.
- Reduced stockouts by 20 to 75 percent.

All of this contributes to improved customer satisfaction and profitability [Lukic, 2012].

Lean Manufacturing and Retail Merchandising Comparison

It is interesting to note that a 2007 study found that Toyota and Seven-Eleven Japan (SEJ) had a similar corporate philosophy of business. Both companies achieve very good business performance, create a unique corporate philosophy, and operate globally around the world. Figure 8.1 gives the different viewpoints of manufacturing and retailing toward the concept of Lean management [Lukic, 2012].

The study found that a Lean philosophy, when supported by information and communication technology, can result in store inventory reduction and increased profitability. It also found that by applying the concept of Lean, retailers develop special relations with suppliers that allow the flow of goods and information to be optimized.

Before looking for ways to use Lean to improve the merchandise planning process, we must first understand that process.

	Lean Management		
	Manufacturing	**Retailing**	
	TPS (Toyota production system)	**General model and concept**	**Seven-eleven key concept**
1	JIT (Just in time)	SCM (Supply chain management)	*CDC (Combination distribution center), NDF (Non-deliverable forward)*
2	Kanban system	DCM (Demand chain management)	Store initiative ordering
3	Production smoothing	SCM (Supply chain management)	Team merchandising
4	Shortened setup time	SCM (Supply chain management)	Customer focus
5	Shortened lead time	Order-delivery	Dominant strategy
6	Standardization of operations	Franchise system	Store initiative ordering
7	Autonomy	In store merchandising	Individual store management
8	Kaizen (Improvement activities)	In store team meeting	Tanpinkanri (item by item control)

Figure 8.1 Production and retail overview of lean management.

Merchandise Management Process

Merchandising is an integral part of retailing and one of the most challenging functions. As a result, it is a major source of potential waste and needs to be both understood and examined carefully.

The major steps in merchandise management entail:

- *Analysis*—This is required because a retailer needs to understand the needs and wants of its target audience.
- *Planning*—It is necessary to plan, since the merchandise to be sold in the future must be bought in advance.
- *Acquisition*—Merchandise to be sold in retail stores needs to be purchased from others (i.e., distributors or manufacturers).
- *Handling*—It is necessary to determine where merchandise is needed and to ensure that the merchandise reaches the required stores at the right time and in the right condition.
- *Control*—As retailing involves spending money for the acquisition of products, it is necessary to control the amount of money spent on buying.

No one in retail can avoid contact with merchandising activities, as they are the day-to-day business of all retailers. As inventory is sold, new stock needs to be purchased, displayed, and sold. As a result, merchandising is often said to be at the core of retail management.

From a more traditional standpoint (seen in Fig. 8.2), the merchandise planning process starts with the buyers forecasting category sales, after which they develop an assortment plan for products in each category and then determine the amount of inventory needed to support the forecasted sales and assortment plan.

Buyers then need to create a plan showing the sales forecast for each month and the inventory (and money) needed to support the sales. Next, the buyers or planners must determine what and how much product to allocate to individual stores. The buyers then need to negotiate with vendors to purchase the merchandise.

On an ongoing basis, buyers monitor sales of merchandise in each category and make adjustments when necessary. This may involve putting merchandise on sale and using the money received to buy merchandise with greater sales upside (or to eliminate some slow-moving stock keeping units (SKUs) in the assortment to increase inventory turns) [Levy and Weitz, 2012].

Before we examine the actual planning process in detail, it is important to understand the factors that affect merchandising.

Factors That Affect Merchandising

Merchandising does not function in isolation (see Fig. 8.3); it is affected by various factors, such as:

- *Size of the retail organization*—The merchandising function varies depending on the size of the retail business. The needs of an independent retailer differ

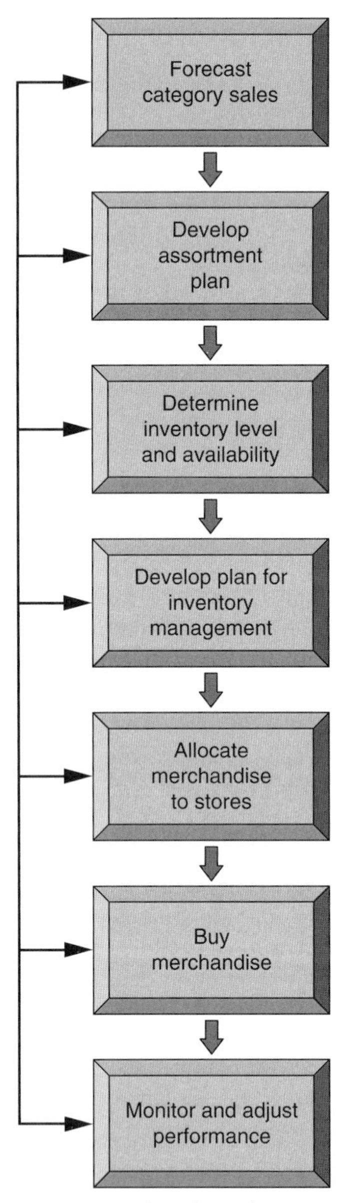

Figure 8.2 Merchandise planning process.

considerably from those of a large chain operation. In the case of a single store, the owner or manager is usually assisted by a salesperson, who may perform the buying function. As the single store grows in terms of business, functional departmentalization may occur, and the number of people involved in the

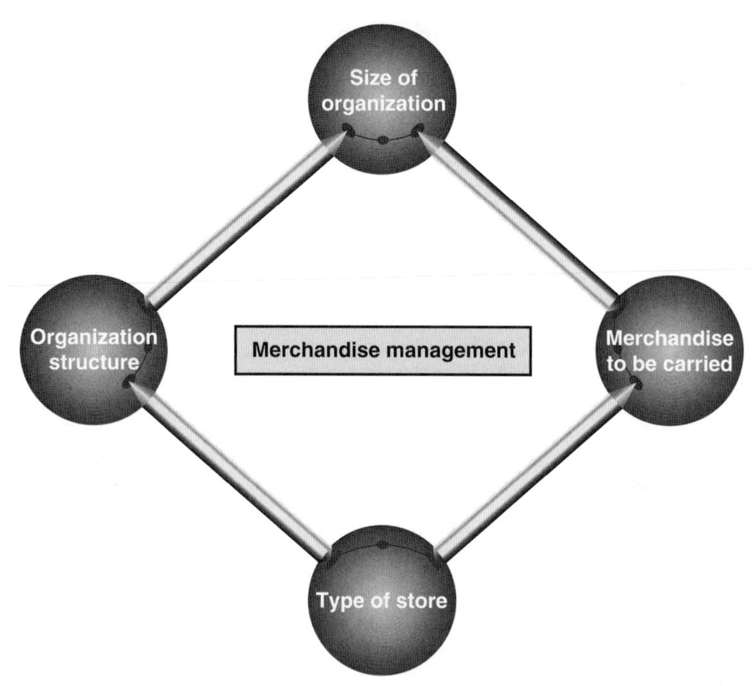

Figure 8.3 Factors that affect merchandise management.

buying process may increase. In the case of a chain store, the buying process may be centralized or decentralized geographically, depending on the organization. A buyer for a chain store may have a fair amount of say in the buying price, as quantities are much larger than those of the single store. As a chain store expands across regions and, at times, across nations, the buyers have to keep regional preferences in mind.

- *Type of store*—The buying function for a mail order catalog, for a direct marketer, or for an e-tailer, would be completely different from that for a brick-and-mortar-only retailer. Mail order buyers need to plan well in advance, as the production of catalogs takes a long time. In addition, the large variety of merchandise sold needs a fair amount of research. Buyers for an e-tail venture need to have a clear understanding of the type of products that consumers will buy on the Internet. In direct marketing or in e-tail ventures, it is often the uniqueness of the product or the competitive price that makes the difference. Thus the nature of the organization is an important factor affecting the function of merchandising.
- *Merchandise to be carried*—The buying for basic merchandise is different from that for fashion merchandise. This is primarily because basic merchandise items (also known as "runners") are always in demand. Examples of basic products are white T-shirts, underwear, and the like. Fashion products are

those that may sell very well in one season or one year, but may not be in demand in the next season or year. Fads are products that have a still shorter life cycle, with sales being very high initially, but with the demand dying down soon. A merchandiser handling fashion products will need to spend more time in the market. It will also need to be aware of the fashion forecasts and trends in international markets.

- *Organization structure*—The organization structure that the retail organization adopts also affects the merchandising function. Some organizations may separate the role of the buyer and the role of the merchandiser. In a smaller organization, one person may carry out all these duties.

Merchandise Manager's Role

Merchandise managers are responsible for particular lines of merchandise. In a department store, there may be separate merchandise managers for different departments: men's wear, women's wear, children's wear, and so on. Each would be in charge of a group of buyers, and the managers' basic duties are typically divided into four areas: planning, directing, coordinating, and controlling, as mentioned before and described in more detail here.

Planning

Merchandise managers formulate the policies for the areas for which they are responsible. Forecasting sales for the forthcoming budget period is required, and this involves estimating consumer demand and the impact of the changes occurring in the retail environment. The sales forecasts are translated into budgets to help the buyers stay within the retailer's financial guidelines.

Analysis is the starting point for merchandise planning. The person who makes the buying decisions for a retail organization must be aware of consumers' needs and wants. A clear understanding of the consumer buying process and what products are actually selling and where they are selling is necessary. Information is obtained from point of sale (POS) data and interaction with the sales staff. External sources of information include surveys, magazines, and trade publications. The information gathered needs to be analyzed, which then forms the basis for the sales forecast. The first stage in merchandise planning is the sales forecast, which we will describe in detail shortly.

Directing

Guiding and training buyers when the need arises is also a function of the merchandise manager. Many times, the buyers need to be advised of additional markdowns being taken on products that may not be selling well in stores. Inspiring commitment and performance on the part of the buyers is a necessary part of the merchandise manager's job.

Coordinating

Usually, merchandise managers supervise the work of more than one buyer. As a result, they must coordinate buying efforts in terms of how well they fit in with the store's image and with the other products being bought by other buyers.

Controlling

Assessing not only the merchandise performance, but also the buyers' performance is a part of the merchandise manager's job. Buyers' performance may be evaluated on the basis of net sales, markup percentages maintained, markdown percentages, gross margin percentages, and stock turns. This is necessary to provide control and maintain high performance results.

Next, we will look at the actual merchandise and assortment planning process.

Forecasting Process

It is important to first understand a typical retail buying organization in order to understand the forecasting process.

At the highest level is the merchandise group, which is managed by a general merchandise manager. Within each merchandise group are departments, which are managed by divisional merchandise managers. Each department has product classifications, which are items targeting the same customer type, such as girls' sizes 4–6, which are followed by a category (e.g., dresses, swimwear, and sportswear categories for girls' sizes 4–6), then finally the individual SKUs, which are the smallest unit available for inventory control (i.e., style, color, and size).

In general, the forecast may be made by the merchandiser based on targets specified by top management. Sales forecasting is the first step in determining the inventory needs for the product or category. The forecasts are typically developed to answer the following questions:

- How much of each product needs to be purchased?
- Should new products be added to the merchandise assortment?
- What price should be charged for the product?

A sales forecast is usually made for a specific period of time, which may be weeks, a season, or a year. The person who forecasts for the product group or category needs to be aware of changes in consumers' tastes and attitudes, the size of the target market, and changes in spending habits.

The highest level of the retail forecasting process is what is known as *category management*, which is the process of managing a retail business with the objective of maximizing the sales and profits of a category. The objective of category management is to maximize the sales and profits of the entire category, not just a particular brand.

However, perhaps the most challenging part of the process is that the retailer has to predict and manage inventory down to the individual style, color, and size of a particular dress that may or may not have been offered for sale before.

Manufacturers find it challenging to forecast demand and create inventory plans for hundreds to maybe even thousands of items. Try forecasting demand for as many as 50,000 items stocked at an individual store on a weekly or even a daily basis.

So I think you can see how challenging the forecasting process can be. However, if one has a good methodology for forecasting along with a good knowledge of forecasting tools and uses available technology, the process can be managed with minimal waste.

Forecasting Steps

The typical forecasting process involves:

1. *Reviewing past sales*—A review of past sales is necessary and helps in establishing a pattern or trend in sales figures. Sales for the past year within a given time period will give an indication of the expected sales in the current year for the same period.
2. *Analyzing the changes in economic conditions*—It is necessary to consider the changes taking place on the economic front, as this has a direct link to consumer's spending patterns. Issues such as economic slowdowns, increases in unemployment levels, and the like can all have an effect on business.
3. *Analyzing changes in the sales potential*—It is necessary to consider any demographic changes in the store's target market that will affect the products to be sold.
4. *Analyzing changes in the marketing strategies of the retail organization and the competition*—The marketing strategy to be adopted by the organization and that of the competition must be considered when forecasting. Questions such as must be asked:
   - Are any new line(s) of merchandise to be introduced?
   - Are any new store(s) to be opened?
   - Are existing store(s) to be renovated?
5. *Creating the sales forecast*—After considering the above-mentioned points, an estimate of the projected adjustment to sales is determined. This information is then applied to the various products and categories to arrive at the projected sales figures.

Forecasting Tools and Methods

No matter what industry you are in, forecasting is a blend of art and science, and, as the saying goes, "All forecasts are wrong" (shorter-term forecasts tend to be more accurate than longer-term forecasts; the same goes for aggregate versus granular forecasts). The idea is to have a process that focuses on minimizing forecast variance and uses a collaborative approach.

It is also very important to use demand history as opposed to sales history where possible in order to have the most accurate forecasts. The problem with sales history is that unless you are 100 percent in stock on all items, you will be forecasting future sales based upon faulty historical data in which sales may be artificially low because products were out of stock or sold later than they were actually demanded (e.g., using "rain checks").

Basically, forecasting is a process of predicting future events and is the underlying basis for all business decisions, including personnel, finance, facilities, and so on.

Types and Horizons of Forecasts

Forecasts are of three general types:

1. *Economic*—These types of forecasts look at general business cycles, including such things as the inflation rate, the money supply, housing starts, and so on.
2. *Technological*—These forecasts are used to predict the rate of technological progress and may affect the development of new products.
3. *Demand*—These forecasts predict the sales of existing products and services.

Forecasts can be for varying lengths of time and at different levels of detail, depending on their purpose. Longer-term forecasts are typically used for facility location and research and development requirements; medium-term forecasts are more for budgeting; while short-term forecasts are for purchasing, scheduling, and job assignments.

The life cycle of a product (i.e., introduction, growth, maturity, and decline) has a great influence on the length of the forecast and the methods used. For instance, introduction and growth stage items require longer-term forecasts than those in the maturity or decline stage. A new product with no previous history will require a more qualitative approach to forecasting rather than a quantitative approach. A mature product may require simpler methods of forecasting, as demand is more static at that point. A product in decline will require keeping a close eye on short-term demand so as not to be left with obsolete inventory.

In retail, there are two general types of merchandise:

1. *Staple or basic merchandise*—As mentioned earlier, these are categories that are always in demand, such as socks and underwear. Generally, they have predictable demand, resulting in the use of primarily historical sales to produce fairly accurate forecasts.
2. *Fashion merchandise*—These are items that are in demand for a relatively short period of time. They are constantly changing (at least the colors and styles change often). Examples of this type of merchandise would be women's apparel and sneakers.

Obviously forecasting for fashion merchandise is much more challenging and riskier than that for basic merchandise, often requiring different methods of forecasting.

Qualitative and Quantitative Approaches

There are two general approaches to forecasting, qualitative and quantitative. Qualitative methods are useful when a situation is vague or there are few or no data existing. This can be the case with new products (including fashion merchandise) or new technology (e.g., tablets or 3-D LED televisions). A qualitative approach involves using a lot of intuition and experience.

Quantitative methods, on the other hand, are used in more stable situations where historical data exist. This method would be used with basic merchandise and items that use current technology (e.g., laptops) and typically utilizes mathematical techniques.

Qualitative Methods The general types of qualitative forecasting are:

- *Jury of executive opinion*—A group of opinions from high-level experts are obtained, sometimes assisted by statistical models.
- *Delphi method*—This method uses a panel of experts and, through an iterative process, continues until a group consensus is reached.
- *Sales force composite*—Estimates from individual salespeople are reviewed for reasonableness, then aggregated.
- *Consumer market survey*—Ask the customer.

In retail, the last two methods are probably used the most. However, in the case of newer or fashion-type merchandise, where previous sales data are not available, retailers can also use market research, fashion and trend services, and vendors as a source of qualitative input for their forecasts.

From a quantitative perspective, for the most part there are what are known as "time series" and "associative" statistical models that use historical demand data to predict the future.

Quantitative Methods The most basic method of quantitative forecasting is known as *time series* and uses a sequence of data points, typically measured at successive points in time spaced at uniform time intervals, to predict future activity. Time series methods make forecasts based solely on historical patterns in the data and typically use time as an independent variable to produce demand forecasts. In a time series, measurements are taken at successive points or over successive periods. The measurements may be taken every hour, day, week, month, or year, or at any other regular (or irregular) interval.

Time Series Models Time series models can contain some or all of the following components (Fig. 8.4):

- *Trend*—A persistent, overall upward or downward pattern with changes caused by population, technology, age, culture, and other such factors. A trend is typically several years in duration.

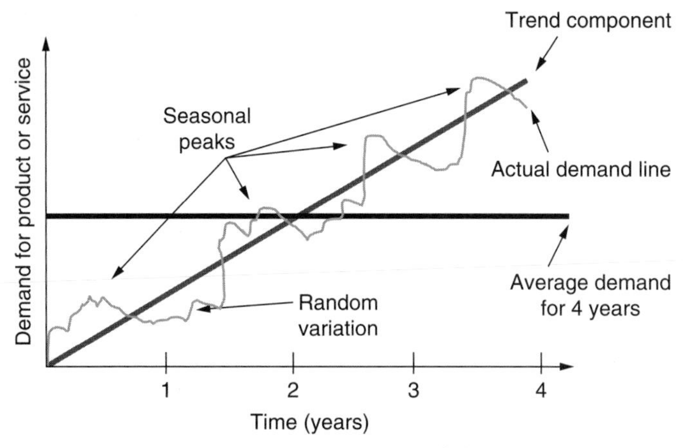

Figure 8.4 Components of demand.

- *Cyclical*—A series of repeating up and down movements that are affected by business cycle, political, and economic factors and are multiple years in duration. There are often causal or associative relationships.
- *Seasonal*—A regular pattern of up and down fluctuations caused by factors such as weather, customs, and the like that occur within a single year.
- *Random*—Erratic, unsystematic "residual" fluctuations caused by random variation or unforeseen events. They are typically of short duration and nonrepeating.

These components may be combined in different ways. It is usually assumed that they are multiplied or added.

The most common quantitative time series models are:

- *Naïve approach*—This approach assumes that demand in the next period will be the same as that in the most recent period (e.g., if January sales were 70, then February sales will be 70). It is simple yet cost-effective and efficient.
- *Moving average*—A series of arithmetic means is used; this method is appropriate if there is little or no trend, as it tends to smooth historical data. Typically, recent data are averaged (e.g., January–March sales are averaged to create an April forecast).
- *Weighted moving average*—This method is used when some trend might be present, as it usually treats older data as less important. The weights used are based on experience and intuition and can minimize the smoothing effect if desired.

For example:

> Weighted moving average forecast for April = 0.7 * March sales + 0.2 * February sales + 0.1 * January sales

In this example, more weight has been given to March demand than to January or February.

- *Exponential smoothing*—A form of weighted moving average in which weights decline exponentially, with the most recent data being weighted the most. It requires a smoothing constant (which ranges from 0–1) that is subjectively chosen.

For example:

New forecast = last period's forecast + 0.8 * (last period's actual demand – last period's forecast)

- In this example, the smoothing constant used, 0.8, will give a relatively high weighting or "smoothing" factor to an over or undersell during the most recent month of history.

Associative Models There are more sophisticated models known as *associative* models, such as linear regression (also known as the least squares method) and multiple regression analysis, which use the relationship of one or more independent variables (x) to predict a dependent variable (y). The reason it is called the least squares method is that the formula draws a "best fit" line through the historical data over time (i.e., the line with the least deviation; see Fig. 8.5). That formula can then be used to predict future values of y.

In linear regression, the relationship is defined as: $y = a + bx$, where a = the y-axis intercept and b = the slope of the regression line.

A simple example of linear regression would be to derive and use this equation to predict future sales (y) by plugging in the number of salespeople (x) that we plan on using. If we know that the two variables are strongly correlated, we can easily

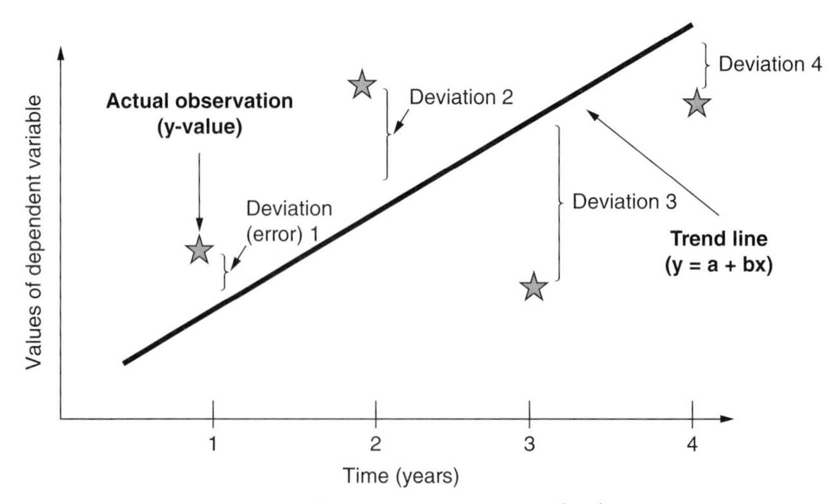

Figure 8.5 Least squares method.

derive the equation using historical sales personnel employment numbers along with historical sales.

To come up with this equation, we must solve for b and then solve for a. The formulas used to derive each are:

$$b = \frac{\Sigma xy - n\bar{x}\bar{y}}{\Sigma x^2 - n\bar{x}^2}$$

$$a = y - bx$$

Once we have solved for a and b, we have our regression formula and can plug in future sales personnel employment estimates to predict future sales.

Seasonality In all the aforementioned time series methods and in linear regression, we can apply what is known as a *seasonality index*. As mentioned, this may reflect actual seasonal sales of an item (e.g., we sell more snow shovels in the winter), or it can be artificially created (e.g., a promotional calendar).

Multiple Regression If more than one independent variable is to be used, linear regression can be extended to multiple regression to accommodate several independent variables (e.g., discounting, promotions, advertising, and other such factors may all have an impact on sales to one degree or another). The formula for this is $y = a + b_1x_1 + b_2x_2 + \ldots$ (similar to the least squares formula, except with multiple independent variables). This is quite complex and is generally done with the help of a computer.

Ultimately, one will want to have the best combination of independent variables for the best possible forecast. A statistic called an "r-squared," which is a measure of the strength of the correlation between y and the various combinations of x's, is calculated. The closer to 1.0 the r-squared is, the better the correlation and, hopefully, the more accurate the forecast.

There are many other statistical methods used, ranging from simple to very complex. The "best-in-class" methods of forecasting use a blend of qualitative and quantitative techniques, including collaboration, both internally with staff from various departments, such as sales, marketing, and finance, and externally with customers and suppliers.

According to a European study of retailers, "Retailer Views on Forecasting Collaboration" [Småros et al., 2004], "most retailers do, indeed, lack dedicated forecasting resources and ... their forecasting tools often are rather simple. The retailers involved in large-scale collaboration are all exceptional in that they have more sophisticated forecasting tools than the other retailers in the sample. In some cases they also have dedicated forecasting personnel. These observations support the notion that intensive Collaborative Planning, Forecasting and Replenishment (CPFR)-style collaboration requires forecasting capabilities that most retailers, even leading ones, currently do not have." They also found that companies that have the most developed automatic store ordering systems and are following a centralized approach to managing store replenishment have invested in forecasting tools and are the companies that seem to be most interested in forecasting collaboration.

However, they also concluded that "companies that lack advanced forecasting processes, tools, and resources can engage in more streamlined collaboration, focusing on information sharing."

The takeaway from this study is that while Collaborative Planning, Forecasting and Replenishment (CPFR) and Quick Response (QR) programs can be complex and costly to implement and manage, some collaboration is better than no collaboration when it comes to forecasting, and simpler may be better (or at least just as good), especially when it comes to small to medium-sized retailers.

As they say, you can't improve a process if you don't measure it, so it is very important to both establish targets and then track and measure forecast accuracy. There are many ways to establish forecast targets, including historical data, contribution, and similar measures. The one I prefer is the *ABC* method, which is a way of classifying items based upon their sales velocity or contribution to profits and can be used not only to set forecasting targets, but also in inventory planning and control, which will be discussed shortly.

To explain the ABC method, one needs to understand a phenomenon known as the *Pareto principle* or *the 80/20 rule*. It states that a relatively small number of your items generate a fairly large percentage of your sales or profits; these are referred to as A items (e.g., Big Macs, Quarter Pounders, fries, and Coke are A items at McDonald's).

In forecasting, these A items need to have more time and effort put into them and typically have better accuracy as a result. The slower movers, known as B and C items, are somewhat less important, require less forecasting time and effort, and typically have more variability (especially the C items, which usually make up the majority of a company's items, yet cumulatively account for only a very small portion of its sales or profits).

Determining the Merchandise Requirement

Once we have generated a reasonable forecast, we need to think about inventory requirements. In merchandising, planning inventory is done at two levels: the creation of the merchandise budget and the assortment plan.

There are two methods of developing the merchandise plan and budget: top-down planning and bottom-up planning.

In top-down planning, upper management works on the sales plan and passes it down to the merchandising team.

In bottom-up planning, the individual department managers work on the estimated sales projections, which are then summarized to arrive at the total sales figures.

In either method, it's important that the detailed forecasts at the bottom add up to the totals at the top (at all times). This is known as having a "one-number" system.

The merchandise budget is the first stage in the planning of merchandise. It is a financial plan, and it indicates how much to invest in product inventories.

The merchandise budget usually comprises five parts:

1. *Sales plan*—How much of each product needs to be sold by each department, division, or store (covered in the previous section).
2. *Stock support plan*—How much inventory or stock is needed to achieve those sales.
3. *Planned reductions*—Actions to take in case a product does not sell.
4. *Planned purchase levels*—The quantity of each product that needs to be procured from the market.
5. *Gross margins*—Profit margins that the department, division, or store will contribute to the overall profitability of the company.

Methods of Inventory Planning

Merchandise planning is a systematic approach taken to maximize inventory return on investment and increase profitability. Proper inventory planning strives to ensure that markdowns caused by overstocks and lost sales caused by stockouts are minimized. There are six general methods (and key performance indicators or KPI) for planning inventory values that will be discussed:

1. Forward weeks of supply (FWOS)
2. Weeks of supply (WOS)
3. Stock-to-sales ratio (SSR)
4. Sell-through (ST) percent
5. Turn
6. Basic stock

Forward Weeks of Supply

Merchandise planning with FWOS is superior for planning appropriate inventory levels at the week level. Using FWOS allows a planner to think about the inventory across time and is essential for managing inventory levels effectively. The goal of effective inventory management is to have enough inventory on hand at any given time to support planned sales until the next delivery arrives. FWOS is calculated as the number of weeks of planned sales from the next week forward that the current inventory value represents. When FWOS is entered into a plan, it will calculate the ending period inventory (EOP) by counting the forward number of weeks of sales and summing the value to calculate the required ending inventory.

The advantage of this method is that inventory is linked to the sales trend across time, so that planned sales trends become the driving factor when using FWOS. Inventory levels are planned to meet future forecasted sales, and the potential for overstock situations is diminished. FWOS puts the focus on maximizing the return on the inventory investment by linking inventory levels directly with planned sales.

A major disadvantage is that FWOS requires planning at the week level, which is not always practical when creating high-level, top-down plans.

Weeks of Supply

WOS is an inventory measure that is calculated by dividing current inventory by *average* sales (rather than future requirements, as in the FWOS method). WOS helps to educate a planner to think of inventory in terms of time.

The main advantage of using WOS is that it's a simple calculation.The disadvantage is that it looks at past trends, not future forecasts, and therefore is not an appropriate measure for calculating inventory needs (especially when there is seasonality involved). By using past sales, WOS determines an inventory position based on where the business has been, not on where the future sales trend is going.

Stock-to-Sales Ratio

SSR is ideal for planning appropriate inventory levels in plans at the month level. SSR forecasts how much inventory is required to achieve the projected sales. SSR represents the proportion of merchandise on hand at the beginning of a period relative to the expected sales for that period. SSR is calculated by dividing stock at the beginning of the period by sales for the period.

In addition to this, gross margin return on inventory investment is also examined as a key metric. It is calculated by dividing the gross margin by the average inventory cost. This is a useful measure, as it helps management see the average amount that the inventory returns above its cost. A ratio higher than 1 means that the company is selling the merchandise for more than it costs the company to procure it.

An advantage of SSR is that inventory is linked to sales values, and it is the most logical key performance measure for planning inventory values in a month-level plan. SSR calculates inventory levels to meet planned sales, resulting in the potential for overstock situations to be diminished.

The disadvantage of SSR is that it looks at only one period of time and does not consider the movement of a sales trend over time.

Sell-Through Percent

ST percent allows a planner to understand the rate at which inventory is consumed as compared to sales. While ST is a KPI, it is better suited for analysis than for planning. The ST percent represents the ratio of sales to beginning period inventory.

It is calculated by dividing sales for a time period by stock at the beginning of the period.

A major advantage of ST is that it links inventory consumption to sales.

A disadvantage of ST is that it looks at only one period of time and does not consider the movement of a sales trend over time.

Turn

Turn, also known as turnover, refers to the number of times during a period that the average inventory is sold and replaced. Turn is a ratio of sales to inventory for a long period of time, usually a season or a year. While turn is the most commonly used KPI, it is better suited for analysis than for planning, since inventory fluctuations over time are flattened. Turn is typically calculated by dividing sales by the average inventory value (the inverse of WOS).

The advantage of turn is that it provides a starting point for planning inventory. Turn targets are typically developed early in the planning process and can be used to roughly estimate inventory levels by month.

The disadvantage of turn is that it uses average values, thus flattening trends.

Basic Stock

The basic stock method of inventory planning calculates a baseline level of inventory that is the same for all months; inventory should not drop below the base level. Planned sales for each month are added to the basic stock to derive the beginning-of-period inventory value. Basic stock value is calculated as average inventory divided by average sales.

The advantage of using the basic stock method is that it supplies a very conservative method of inventory planning.

The basic stock method of inventory planning is an option to consider for businesses with very consistent sales and inventory levels, meaning that there is little seasonality or fluctuation in sales, as the baseline stock is the same for all months.

A disadvantage is that it uses average values, thus flattening trends, and is not an appropriate method to use to calculate planned inventory for seasonal businesses, emerging categories, or products with less predictable selling patterns. It can be well suited for planning at the lowest level of detail (SKU).

Summary for Determining Merchandise Requirements

In general, FWOS is a good method for calculating planned inventory levels and should be used in all plans containing the week level of time. It is the best method of planning inventory to support the projected sales trend over time. FWOS is a retail industry best practice inventory planning measure. The FWOS "target" is often set by using the ABC method of inventory management. Targets are typically set by an ABC code. The A items, which are faster movers and contribute more to sales and/or profits, typically have lower FWOS targets, as retailers and wholesalers carry the bulk of their inventory in these items. The B and especially the C items may have very high FWOS targets, but the actual inventory quantity may not amount to much, as they are slower movers.

SSR is the most appropriate method for high-level, top-down inventory plans and other inventory plans that do not contain the week level of time. SSR forecasts the appropriate inventory level required to support projected sales. SSR is a leading retail industry planning measure [Parker Avery Group, 2013].

Open-to-Buy for Fashion Merchandise

Another method (similar to FWOS) that retailers use, especially for fashion merchandise planning, is "open-to-buy" (OTB), which can be calculated in either units or dollars. OTB is the difference between how much inventory is needed and how much is actually available. This includes inventory on hand, inventory in transit, and any outstanding orders.

In order to take advantage of special buys or to add new products, some of the OTB dollars should be held back. This also allows the retailer to react to fast-selling items and restock the shelves quickly.

A retailer should consider maintaining an OTB plan for its business as a whole, but should also plan for each category of merchandise that it stocks. The plan can be maintained on paper, in a spreadsheet, or by purchasing one of the several retail software packages available that contain OTB programs.

The OTB formula is as follows:

OTB (retail) = planned sales + planned markdowns + planned end-of-month inventory − planned beginning-of-month inventory

For example, a retailer has an inventory level of $150,000 on July 1 and a planned $152,000 end-of-month inventory for July 31. The planned sales for the store are $48,000, with $750 in planned markdowns. Therefore, the retailer has $50,750 OTB at retail.

Note: Multiply that number by the initial markup to reach the OTB at cost. If the markup is 60 percent, then the OTB at cost is $20,300.

As a replenishment tool, OTB is not appropriate for all categories of merchandise (as mentioned earlier). It is most appropriate for fashion merchandise, where the specific items may change, but the departments, classifications, and subclassifications remain relatively stable, and for seasonal merchandise, where inventories are brought in at the beginning of the selling season and need to be managed down to predetermined ending levels at the end of the selling season.

OTB is also not appropriate as a replenishment tool for day-in and day-out basics. These staple items are replenished more effectively using an automatic replenishment program that runs off of predetermined minimum and maximum inventory parameters. In the case of these in-stock basics, OTB may still serve a valuable budget and control function at a department or category level.

Using OTB requires some investment in time and attention to build a realistic plan, and discipline to maintain it through the year or a season. However, it can yield huge results quickly in most situations, from increased sales to leaner inventories and reduced markdowns and overstocks.

Economic Order Quantity and Reorder Point models

Another relatively easy method for answering the question of "how much" to buy or order is the economic order quantity (EOQ) model, which is especially useful to the small retailer.

The simple EOQ calculates the optimal number of units that a company should add to inventory with each order to minimize the total costs of inventory. The calculation balances total inventory holding or carrying costs (i.e., cost of capital, taxes, storage, damage, obsolescence, and so on) against ordering costs. There are more complex versions of the model that factor in quantity discounts.

The concept is that a larger order size will have greater holding costs, but lower ordering costs, as the orders are made less often (and vice versa). The EOQ calculates the trade-offs between holding cost and ordering cost to find the optimal order quantity with the lowest total cost.

A company's holding or carrying costs can range from 15 to 40 percent annually, so the size and frequency of orders can have a significant impact on the bottom line. The EOQ is typically best used for basic as opposed to fashion merchandise.

The simple EOQ formula is as follows:

EOQ = square root of [(2 × annual demand in units × order cost) /annual holding costs per unit]

So, for example, if a business sells 1000 iPhones per year, has ordering costs of $10 per order, and has holding costs of $0.50 per unit per year, the EOQ would be 200 units (with orders placed 5 times per year).

While the EOQ answers the "how much" question, the reorder point (ROP) answers the "when" question.

There are two general types of ROP models, the Q (quantity) and the P (periodic review or fixed period) model.

The Q model multiplies the daily average demand by the lead time to determine the ROP. So, for example, if daily demand is 10 units and replenishment lead time is 3 days, the ROP will occur when the inventory drops to 30 units. Uncertainty in demand and lead time can be dealt with by holding additional safety stock. The safety stock can be statistically calculated based upon a desired customer service level for that item (higher for A items, lower for C items), or a rule of thumb can be used, such as average demand during half of an item's lead time (in the previous example, that would be 15 units additional safety stock).

The P model bases reorder quantity upon a planned periodic review of an item and an inventory target (usually in terms of days of supply). For example, faster-moving A items may be reviewed daily and, as they are high-volume items with more accurate forecasts, have lower days of supply targets. Slower-moving C items, on the other hand, may be reviewed only biweekly and have greater days of supply targets because of their smaller sales quantities and greater demand variability.

As a result, the reorder quantity in the P model will constantly be changing, as it is based upon a targeted number of days' supply of future demand, rather than the fixed amount used in the Q model.

Acquisition

Once the analysis and planning process is completed, the acquisition or buying process begins (handling and control processes are, in general, covered in the next two chapters on store operations and distribution operations).

In the retail industry, the acquisition process is primarily handled by a buyer, who selects what items will be stocked in a store, based on a forecast using a process similar to what we just described.

The retail buyer tends to work closely with designers and sales representatives, and to attend trade fairs and visit wholesale showrooms and fashion shows to understand short- and long-term trends. In the case of a smaller retailer, the buyer may have broader responsibilities, including sales and promotion, as opposed to a larger retail chain, where there are more levels of management, including trainee buyers, assistant buyers, senior buyers, buying managers, and buying directors.

A buyer, depending on his or her level in the organization, can have roles and responsibilities that include a wide array of activities ranging from buying and merchandising to other areas such as merchandise presentation, advertising, shortage control, and staff development and training.

Lean Buying

We will now discuss buying from a Lean perspective while focusing on the negotiation process with vendors with regard to prices and margins, favorable terms, discounts, and transportation allowances.

In general, a retailer will either buy product from a national brand manufacturer (or wholesaler) or develop its own private-label brands (discussed in more detail in Chap. 13). In the case of buying products, they are usually bought either from existing "regular" suppliers or from new suppliers that the retailer has never used before.

The types of suppliers besides direct from the manufacturer include full-service wholesalers that perform many services for retailers, including shipping, storing, credit, and information; limited-service wholesalers that provide less in the way of retail services but have lower prices; and agents and brokers, who don't take title to the goods, but instead act as intermediaries.

The actual buying decision may be made as often as five or six times per year for fashion items and less frequently for staple merchandise.

The process of buying national brands will typically include meeting with vendors to discuss the vendor's merchandise performance during the previous season, reviewing the vendor's merchandise offering for the coming season, and then returning to their offices to discuss this information with the buying team before negotiating with vendors.

Negotiation

Negotiating with vendors involves two-way communication designed to reach an agreement when the two parties have both shared and conflicting interests.

Negotiation Issues

Negotiation issues in retail may include [Levy and Weitz, 2012]:

- Maximizing price and gross margin, as we all want to buy at the lowest price in order to have the highest margin. Some factors that help in this are:
  - Margin guarantees—The vendor guarantees a specific sales price. If a product has to be marked down, the vendor agrees to give the retailer markdown money.
  - Slotting allowances—Retailers may need to negotiate slotting fees or allowances, which are charges to stock a new item (viewed by many vendors as extortion).
- Additional markup opportunities, such as excess inventory from the manufacturer.
- Purchase terms modified to stretch out payments to the vendor, thereby improving the retailer's cash flow.
- Buying merchandise that is exclusive to the retailer, so that it can differentiate itself.
- Advertising allowances offered by the supplier to help defray some of the retailer's promotional costs.
- Transportation issues, which can include who pays the transportation as well as how much.

Negotiation Tips

There are many tips that help in a negotiation process. Some of the more important ones are [Waters, 2013]:

1. *Be prepared*—Maybe the most important thing that a retailer (or anyone) can do is to be prepared and informed when going into negotiations. Learn as much about the supplier and its products as possible, such as its prices compared to those of the competition and the level of service it provides to its customers. Always set goals to know what you want and what you'll settle for.
2. *Be honest*—Bluffing or lying can do more damage than good in the vendor negotiation process. Lying not only is unethical, but can be difficult to maintain. However, be careful not to give away your bargaining power, as you don't need to reveal everything during a negotiation.
3. *Show your business's potential*—When you are meeting with a potential vendor for the first time, it's a good idea to start the negotiation with some history about your retail business, including any future plans.
4. *Incentives*—In order to receive the best price, payment terms, advertising allowances, and even exclusivity from a vendor, you have to ask for it. Ask what incentives you qualify for and let the negotiations begin from there.
5. *Competition*—You can bring up the vendor's competition in the negotiation process without disclosing any pricing or other confidential details.
6. *Compromise*—Just like the retailer, the vendor must make a profit. The relationship should be looked at as a collaborative process, not a "beat down." So always try to consider the outcome for the supplier.
7. *Consider the long term*—Having a solid, trusting relationship with a supplier will help your business. You may even get more incentives from the vendor to maintain a long-term relationship.
8. *Don't rush the process*—Salespeople tend to use pressure to close a deal. If this is the case, feel free to ask for time to consider and discuss the offer.

Processing the Order

Many medium and large retailers use computers to complete and process orders (often using electronic data interchange (EDI) and based upon QR inventory planning), where each purchase is entered into a computer.

Smaller retailers write up and process orders manually, but thanks to innovations in technology in terms of functionality, accessibility, and affordability, they may instead be able to place orders electronically.

As mentioned earlier in the book, EDI (and Extension Markup Language (XML) via the Internet) can reduce data entry errors and costs and also create other collaborative opportunities to increase sales and develop a long-term partnership with customers and suppliers.

Collaboration in Product Development and Design

As with the supply chain in general, retailers are looking for longer-term relationships with partners that involve collaboration, cooperation, and trust in order to develop a "win-win" situation.

Examples of this include:

- Best Buy and Toshiba combined their shopper insights to create the first child-centric laptop, the Satellite L635 Kids' PC, with unique attributes such as a fingerprint-resistant screen, a wipeable keyboard, and an Internet browser that was kid-relevant and secure [CNET, 2010].
- Walmart has partnered with GE to develop and market licensed GE-branded kitchen small appliances [Wal-Mart, 2010].
- Starbucks embraced a form of "open innovation" with the launch of its mystarbucksidea.com website, through which it has received more than 100,000 ideas so far for new products and services ranging from digital rewards to new coffee flavors to the little extras, like splash sticks to keep clothes clean [Business Wire, 2013].

While these are great examples of the acquisition process not being viewed as just about price, we always need to look at the process as a whole, so that we can identify opportunities to reduce our costs and improve efficiency while still ending up with a "win-win" situation that both parties can be happy about.

Merchandise and Assortment Planning Software

I have always believed that technology can enable only a good process. Going hand-in-hand with a Lean retail organization, one needs a good merchandise and assortment planning software system. Luckily, advances in technology have made these types of systems affordable to retailers of all sizes, as:

> Underlying every well-run retailer are the business processes and infrastructure necessary to make sure customers get the product they want, at a price they can afford, when and where they want it. Orchestrating an item from initial creation through to its final clearance means managing tremendous amounts of information such as costs, deals, prices, orders, returns and more. In many cases, this information is stored or scattered across multiple systems, making end-to-end visibility and efficient execution challenging. An integrated, centralized repository for all merchandising transaction data provides retailers with the information they need to make better-informed decisions based on a single version of the truth.

Retailers need to make the connections between items, locations and suppliers, track purchase orders, monitor deal income, manage replenishment settings, execute pricing decisions, and aggregate transaction information into stock ledger reporting levels. As the central source of all information, merchandising solutions provide organizations with an accurate view of perpetual inventory and financial performance across their entire retail organization [Oracle, 2013].

These systems may have a variety of functionality, including:

1. *Merchandise financial planning*—Used to manage financial targets and open-to-buy budgets.

2. *Assortment and space planning*—Automates the assortment planning process, creating more flexibility in planning in-depth assortments.
3. *Price and markdown planning*—Plans, forecasts, and analyzes campaigns, promotions, and events.
4. *Allocation and replenishment*—Improves order accuracy by helping to determine optimal demand-based allocation and replenishment needs.

Distribution Requirements Planning for Allocation and Replenishment

There is a particular software tool used by many manufacturers and distributors, known as distribution requirements planning (DRP; see Fig. 8.6), that deserves special mention. When used in conjunction with forecasting software, it has shown real promise in the allocation and replenishment process at the store level.

DRP calculates finished goods replenishment based on forecasted and actual demand at the end of the distribution system and is accumulated backward, netting inventory at each level to determine replenishment requirements at the subsequent level (which can be either a distribution center (DC) or a production site).

There is a recent movement to use DRP as a replacement for current systems, which are more push in nature and base replenishments upon past shipments from the DC to the stores, ignoring current plans and events at stores.

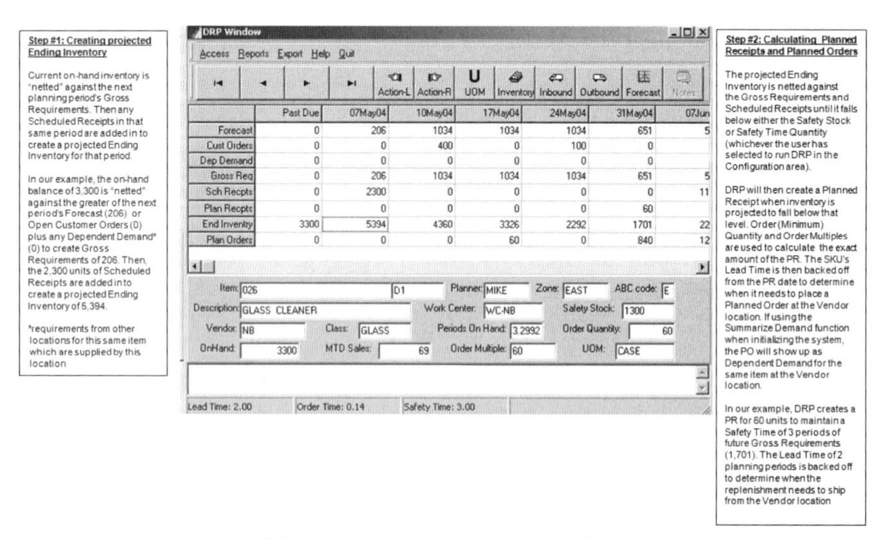

Figure 8.6 Sample DRP screen and description. (From *PSI planner* for windows™, copyright © 1998–2013.)

According to a presentation given at the GS1 Connect 2012 meeting [Smith, 2012], some large retailers, in conjunction with their key suppliers, have implemented store-level DRP with great success, such as:

- Sony Canada (at its company stores and key retailers):
  - Improved its store-level forecast accuracy by 30 percent.
  - In-stock levels went from 87 to more than 95 percent.
  - Supply chain inventory was reduced by 20 percent.
- Lowes and Black & Decker:
  - Fill rate improved to 98 percent on a consistent basis.
  - In-stock levels went from 92 to 98 percent.
  - Forecast accuracy improved by 10 percent
  - There was a significant reduction in inventory.
- Sam's Club and Kraft:
  - Improved order forecasting by 7.7 percent.
  - In-stock levels increased.
  - Supply chain inventory decreased.
  - Lower warehousing, transportation, and obsolescence costs.

This "family" of merchandise and assortment planning software tools can almost be considered a requirement for implementing any of the collaborative tools mentioned in Chap. 4, such as QR, Efficient Consumer Response (ECR), CPFR, and vendor-managed inventory (VMI), as the number of SKUs that retailers offer continues to expand and the retail supply chain becomes more complex.

At the end of the day, they are a wise investment to enable a Lean merchandise planning process by improving collaboration and visibility both internally and externally with suppliers, partners and customers.

Wholesale Forecasting and Inventory Planning

The previously mentioned forecasting and inventory planning methods and software (in some cases specific to wholesale or to warehousing) apply to the wholesale distribution business as well.

Like retail, the wholesale distribution industry is faced with shrinking margins, growing costs, new competitors, and demanding customers. These days, many wholesalers offer value-added services and have introduced private-label goods to increase their margins and profitability. In the distribution industry, the one core asset is inventory. If you have too much of it, you waste resources and risk obsolescence. If you have too little, you may lose customers. Effective supply chain management is what can separate a successful distribution company from those that struggle and often fail.

Additionally, smart wholesalers, like manufacturers, collaborate with their customers using programs such as VMI, ECR, and CPFR to improve the accuracy of their forecasts, reduce inventories, and improve service levels.

Wholesalers can use these methods and technology to:

- ▲ Increase sales through better in-stock positions.
- ▲ Improve buyer productivity and ability to manage more items.
- ▲ Increase inventory turns while ensuring geographic assortments.
- ▲ Model complicated timing and pricing variables to optimize buys.
- ▲ Optimize purchasing, load building, and vendor grouping efficiently.
- ▲ Employ dynamic time-phased inventory policies.
- ▲ Adjust forecasts automatically as demand changes.
- ▲ Free up working capital for more strategic uses.

Once retailers and wholesalers have a Lean process and the information to manage it, they still need a way for executives to monitor and control it. That's where sales and operations planning (S&OP) comes in.

Sales and Operations Planning (Also Known as Integrated Business Planning) in Retail and Wholesale

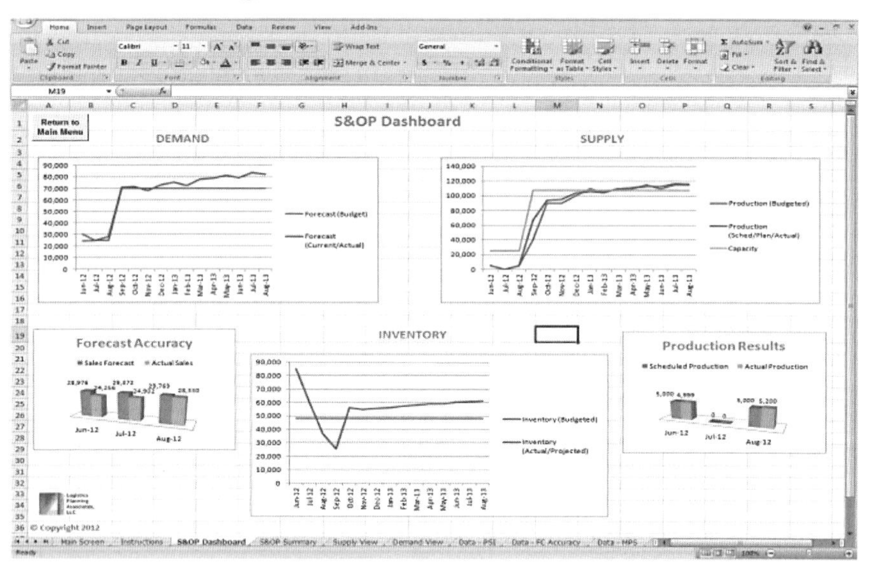

In its simplest terms, S&OP is a way for a business to ensure that supply can match demand, at least in the aggregate. S&OP has been around since the 1990s and is used primarily in manufacturing. In recent years, S&OP has been expanded to include the technologies, applications, and processes that connect the planning function across an enterprise to improve organizational alignment and financial performance.

According to Ronald Ireland and Mary Adamy of Oliver Wight Americas, Inc.,

> The retail sector has been slow to adopt S&OP however, and there are limited instances of retailers operating S&OP best practices ... and

retailers not utilizing S&OP hinders the supply chain from operating effectively in two primary ways: 1) the retail supply chain network remains unpredictable because changes are not anticipated with enough advance notice for the supply chain to most efficiently respond; 2) the benefits from S&OP are not realized by the retailers, and the benefits that could be realized by retailer suppliers are not maximized. The benefits include simultaneous improvements in product availability and fulfillment, while at the same time reducing inventory investment and increasing sales volume [Ireland and Adamy, 2010].

Lean teams plan and execute on an operations level, but S&OP can be a great tool to make the connection between the kaizen event goals and objectives and the corporate ones. As we know, inventory is one of the eight wastes and covers variability in a process. Through the use of S&OP, inventory can be directly controlled. Overall, there are two general ways to reduce inventory: (1) more accurate forecasts and (2) shorter cycle times. The S&OP process attempts to improve and control both of these.

S&OP Defined

S&OP, also called aggregate planning, is a process in which executive-level management regularly meets and reviews projections for demand, supply, and the resulting financial impact. S&OP is a decision-making process that makes certain that tactical plans in every business area coincide with the company's business plan. The net result of the S&OP process is that a single operating plan is created that allocates company resources.

S&OP increases teamwork among departments and helps to align your operational plan with your strategic plan. It is a process in which various targets are set (for example, forecast accuracy, inventory turns, and the like), and progress against the strategic and operational plans is reviewed in a series of meetings.

The objective of S&OP is to have consensus on a single operating plan that meets forecasted demand while minimizing cost over the planning period. It should allocate people, capacity, materials, and time at the least possible cost, while ensuring the highest customer service possible [Myerson, 2012].

The S&OP Process

The series of meetings prior to the final S&OP executive-level meeting (Step 5 in Fig. 8.7) is:

1. *Data gathering*—Data such as the most recent month's actual sales and production or purchases are gathered to update various models and generate updated demand and supply requirements.
2. *Demand planning*—A demand planning cross-functional meeting is held, at which forecasts are reviewed with a team that includes (at the very least) operations, sales, marketing, and finance. Typically, forecasts have been generated

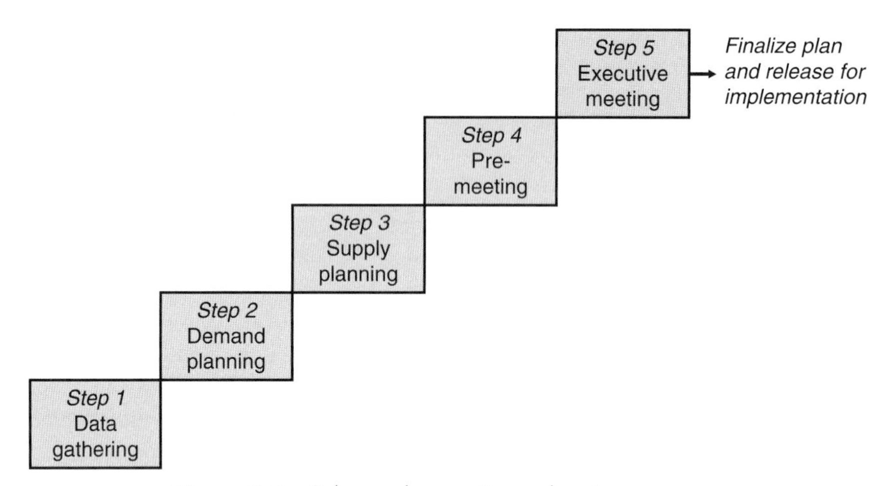

Figure 8.7 Sales and operations planning process.

statistically and aggregated in a format that everyone can understand and confirm (e.g., sales might want to see forecasts and history by customer in sales dollars).

3. *Supply planning*—A supply planning cross-functional meeting is held at which agreed-upon forecasts have been "netted" against current on-hand inventory levels to create production or purchasing plans. Again, these data will usually be reviewed in the aggregate by product family in units, for example.
4. *Pre-meeting*—A pre-S&OP meeting is held, at which data from the first demand and supply meetings are reviewed by department heads to ensure that consensus has been reached [Myerson, 2012].

S&OP in Retail Versus Manufacturing

The S&OP processes for manufacturing and retail are very similar. The main differences are that the sponsors and titles of each step and the details of each review, such as issues, data, and decisions are different.

In the case of retail, the executive participants and responsibilities in the executive S&OP session would be:

- *S&OP process owner (CEO or president)*—This person ensures that the decisions are made and carried out.
- *Finance (CFO)*—This person presents the final financial plan and discusses financial issues.
- *Supply (VP of supply chain)*—This person presents the supply plan and identifies imbalances in supply and demand.
- *Demand (VP of replenishment or POS forecasting)*—This person presents the demand plan and discusses market and customer issues.

- *Product management (VP of merchandising)*—This person discusses category management and identifies new product launches and assortment planning.

S&OP Benefits

What retailers and wholesalers using an S&OP process realize is that it helps them to better align with their suppliers over a longer planning horizon, which can minimize volatility and take advantage of possible opportunities. This aligns with the manufacturing trend of collaboration with retailers that we previously discussed, such as QR and CPFR, and can be a point of intersection between the supplier's and the retailer or wholesaler's S&OP process.

There are many benefits to using an S&OP process, such as breaking down artificial "silos" between departments in an organization, better communication, increased transparency, more reliable plans, less firefighting, and the development of common goals and plans that benefit the entire company, as opposed to focusing on (and suboptimizing) functional goals [Ireland and Adamy, 2010].

In summary, when a retailer develops an organizational structure with one owner of S&OP across geographies and channels, it can use S&OP to help balance open capacity across merchandising, finance, and the supply chain, while improving speed and responsiveness across the organization.

The real heart of a Lean, customer-focused retail organization is seen in its operations, which we will look at next.

CHAPTER 9

Being Lean: Store Operations Management

In this chapter, we will examine the basics of retail store management and where Lean concepts can be applied for process improvement. To accomplish this, it's best to break store management into the logical areas of (1) operations, (2) layout, and (3) customer service.

We will also discuss possible Lean applications in distribution operations that apply to both the retail and wholesale sectors (and will go into more detail on this in the next chapter).

Recent Trends

In recent years, Lean, Six Sigma, and now their hybrid, Lean Six Sigma, are being applied by some larger organizations in the retail sector. This is a result of innovative retailers starting to understand that by using Lean principles, they can develop and implement strategies that will increase performance and drive profitability for their businesses.

One example is Topaz, Ireland's largest fuel and convenience brand, which has developed a Lean vision for its business to create a retail operation that is faster, better, and more convenient for its customers. It has tried to create a continuous improvement culture by enabling its frontline employees to show it how they can work smarter for the firm's customers.

Topaz has started this journey in three phases:

1. *Identify and eliminate waste*—Teach employees to recognize waste, ask them what's not working, and have them design and implement solutions.
2. *Management by exception*—Develop key performance indicators (KPIs) that apply to all levels. Measure and report results to employees and respond quickly to improve exceptions.

3. *Continuous improvement*—Establish a culture of continuous improvement in which results are reviewed during daily team briefings and targets are set locally, with good performance rewarded.

   In Topaz's case, the results have been impressive:

- *Sales*: +5.0 percent Topaz sales versus the rest of the forecourt market in 2009.
- *Service*: 90.7 percent Gapbuster ranked Topaz in the top 2 percent worldwide for mystery shoppers.
- *Productivity*: –25.6 percent actual labor cost savings in two years.
- *Margins*: –0.8 percent reduction in shop shrinkage from –1.6 percent to –0.8 percent.
- *Cash flow*: –4 days reduced stockholding from 15 days to 11 days

[Gleeson and Landy, 2010].

> Margins in the retail industry are extremely tight, so every penny counts. Lean Six Sigma is a way to identify opportunities to eliminate waste, unwanted variation, and errors in retail business processes, all of which cost money, add no value for customers, and don't contribute to the bottom line.
>
> Although Lean Six Sigma's roots may be in manufacturing, it has been applied in a wide variety of other industries and businesses. If there is a repetitive process involved, the approach can be used to improve it. A retail chain serving millions of customers from hundreds of stores has many such processes. Lean Six Sigma helps first by getting people to understand their business from a process perspective. Most managers who are trained in traditional business schools are taught to view the business as a set of functions, such as marketing, purchasing, human resources, etc. But customer value is created by processes, not functions. When processes are examined with Lean Six Sigma, opportunities for improvement often jump out from the analysis. For example, process maps may reveal a great deal of duplication of effort, or important tasks for which no one has clear responsibility, or confusing lines of authority. Lean Six Sigma's super-effective DMAIC (Define, Measure, Analyze, Implement, and Control) project execution framework provides a way to pursue these opportunities [Pyzdek, 2013].

Kum & Go convenience stores, an Iowa-based chain with more than 400 locations in 11 states, knows that the convenience store market is highly competitive. The retailer has taken to using Lean Six Sigma to help it compete and has practiced Lean Six Sigma for several years. The firm thinks this is a way to deliver a great retail experience to customers, and also to differentiate itself from the competition [Pyzdek, 2013].

Store Operations

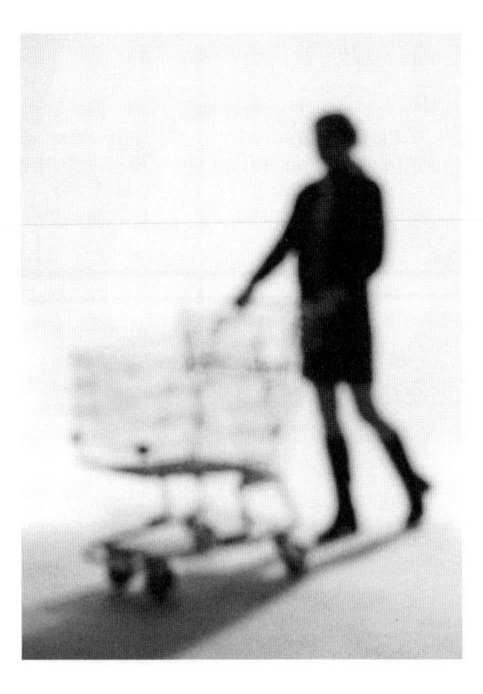

It has become harder for retailers to differentiate themselves based upon products, since most competitors carry similar national brands. As a result, the idea of using store operations as a competitive advantage has become increasingly important.

Service Matrix

When considering the ideal process for your business, a service process matrix is also a very useful tool. It is a classification matrix of service industry firms based on the characteristics of the individual firm's service processes (see Fig. 9.1).

Your company's position in the service process matrix may present some challenges for management. For firms with lower labor intensity, plant and equipment choices would be very important, indicating a need to monitor technological advances. As capacity may be somewhat inflexible because of equipment capabilities, scheduling service delivery is more important, so demand must be managed.

For companies with high labor intensity, workforce issues such as hiring, training, employee development and control, employee welfare, and workforce

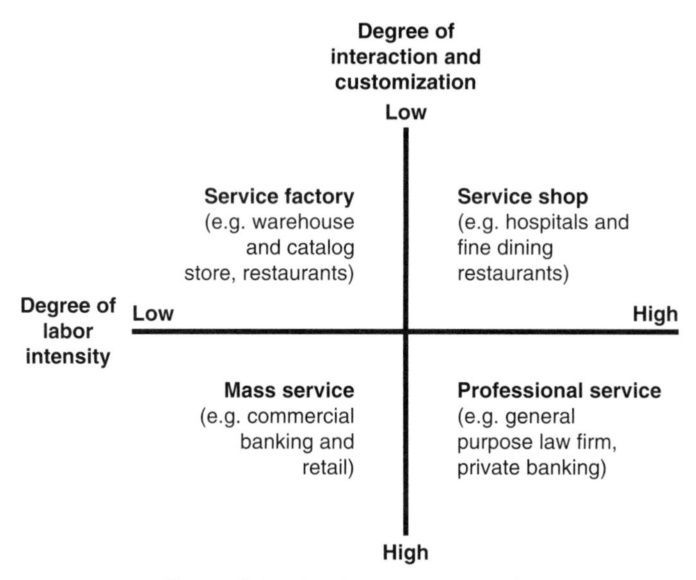

Figure 9.1 Service process matrix.

scheduling are extremely important. In contrast, an organization with low customer interaction and customization would confront more marketing-oriented challenges than others.

Companies with a high degree of interaction and customization need to manage higher costs. In addition, more highly skilled labor costs more and requires more attention, a better quality of work life, and better benefits. The managerial hierarchy tends to be flatter and less rigid.

Service Blueprint

A good place to start when thinking about retail operations improvement is how your operation interacts with the customer. A great tool for mapping this process is known as a "service or operations blueprint." Service blueprinting focuses on the interaction between the customer and the provider. It defines three levels of interaction, with each level having different management issues. It helps to identify potential failure points (see Fig. 9.2).

Some interaction with the customer is necessary, but this can affect performance negatively. The better these interactions are anticipated in the process design, the more efficient and effective the process will be. It is important to find the right combination of cost and customer interaction for your business.

A large retailer may have multiple blueprints for areas such as store maintenance, inventory management, credit management, and store displays. When the

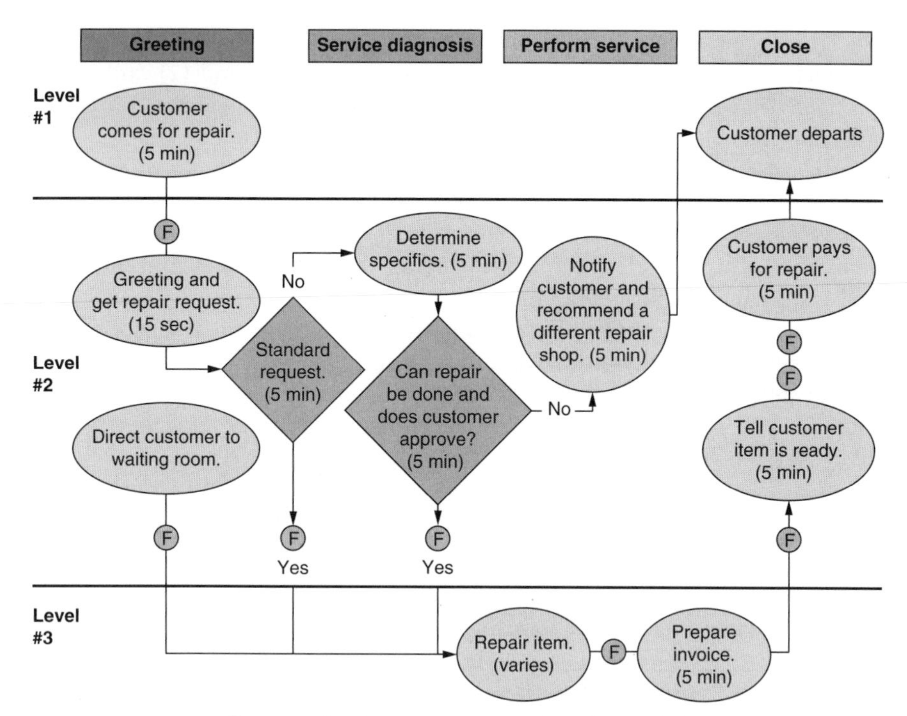

Figure 9.2 Service or operations blueprint.

retailer makes changes to its format or its standard operating procedures (SOPs), it must also adjust the operations blueprint.

Management Responsibility

Levy and Weitz, in their book *Retailing Management* (McGraw-Hill Irwin, 2012), have identified four areas of responsibility that store managers directly affect, each of which can be used for a competitive advantage and each of which contains non-value-added activities. They are cost control, managing employees, managing merchandise, and customer service. We will now discuss each of those areas from a Lean perspective (except for customer service, which will be covered separately at the end of this chapter).

Cost Control

One of a store manager's key responsibilities in running a retail operation is to manage labor productivity, which has a direct impact on operating costs.

Productivity is the ratio of outputs (goods and services) to inputs (resources such as labor and capital). The goal, therefore, is to improve productivity by measuring process improvement through the relationship of outputs to inputs. Only through productivity increases can our standard of living improve.

There are two general types of productivity measurements: (1) single-factor productivity, which typically uses only labor as an input, and (2) multifactor productivity, which includes not only labor but also material, energy, capital, and miscellaneous inputs.

In general, labor contributes around 10 percent directly to productivity improvements, and variables such as basic education appropriate for the labor force, the diet of the labor force, and social overhead that makes labor available can have an impact. Capital investment (including investments made in technology) and management contribute to the majority of productivity improvements, both of which require employee training and participation in order to be successful.

In the case of retail, which is typically labor-intensive, we know that productivity is frequently focused on unique individual attributes or desires, and therefore it is often an intellectual task performed by professionals that can be difficult to mechanize and difficult to evaluate for quality.

Retail Productivity (and Process) Improvement Examples

Starbucks is always looking at ways to improve labor and equipment productivity at its stores and has had a team of 10 analysts continually looking for ways to shave time. Some improvements the company has made are:

Productivity Improvement	Benefit
Stop requiring signatures on credit card purchases under $25	Saved 8 seconds per transaction
Change the size of the ice scoop	Saved 14 seconds per drink
New espresso machines	Saved 12 seconds per shot

As a result, operations improvements have helped Starbucks increase yearly revenue per outlet by $200,000 to $940,000 in six years. Productivity has improved by 27 percent, or about 4.5 percent per year.

At another retailer, Taco Bell, management looked to improve productivity and made the following improvements:

- Revised the menu.
- Designed meals for easy preparation.
- Shifted some preparation to suppliers.
- Set up an efficient layout and automation.
- Increased training and employee empowerment.
- Adopted new water- and energy-saving grills.

The results of these improvements were:

- Preparation time cut to 8 seconds.
- Management span of control increased from 5 to 30.
- In-store labor cut by 15 hours per day.

- Stores handle twice the volume with half the labor.
- Conserve 300 million gallons of water and 200 million kWh of electricity each year, saving $17 million annually.

[Heizer and Render, 2013]

Retail Layout and Workplace Organization Examples

The store format, size, and allocation of space can also be critical to productivity. Most large retailers use what are known as "prototype" stores. That is, they conform to a standard format in terms of construction, layout, and operating procedures. Operations are performed the same way in all retail locations.

Some stores have opened smaller-format stores as well to reduce costs and improve the bottom line. An example of that is Toys"R"Us, which during the holiday season opens "pop-up" stores in malls. These stores are small in size and temporary. Other examples of retailers opening smaller-format stores include Home Depot, Tiffany's, and Tesco.

According to an article in *Produce News* [Pelger, 2013],

> Grocery chain retailers are even welcoming the change to a smaller store format with open arms. They want and need to turn over a new slate by creating a different niche. Downsizing to a less massive store will bring essential savings to the bottom line.
>
> The average supermarket store is close to 50,000 square feet in size, and it takes a whole lot of energy to heat the building in the winter and cool it in the summer. Add a multitude of lighting and refrigeration necessities, and the utility bills rocket into outer space. With consumers cutting back on their food-spending budgets, those enormous energy costs will take longer to pay off.
>
> Smaller stores also do not require as much inventory stock amounts. Only the basic items furnished on grocery shelves and in the fresh departments that consumers prefer the most are required. Therefore, no unnecessary assets of slow-moving items tie up space and money.
>
> Greater productivity can be found in labor by operating smaller units. Stocking levels of a faster-selling, limited line of products makes sales easier to handle than when every item is made available. Maintaining and cleaning small stores takes less time.

So, in effect, the smaller format allows retailers to become Leaner and at the same time follow a trend for consumers who want to shop in smaller stores, according to a TNS Retail Forward survey referred to in the *Produce News* article.

The Lean concept of layout is critical here, and there are many ways to use it to improve store space productivity, such as using vertical displays that occupy less room and can hang on walls or from ceilings. Space that was not previously used now may have point of sale (POS) displays and vending machines.

As we discussed in Chap. 5, 5S or workplace organization improves both productivity and safety. As a result, many retailers have implemented 5S.

Sonae MC, a global leader in the retail sector operating supermarkets and hypermarkets, started its Lean journey with 5S "to give workers a structure for eliminating disorder in physical environments, such as stocking and receiving areas. The next area was visual management—the implementation of labeling systems, markings, color codes, and other visual tools to provide visual guidelines for an orderly environment. Finally, there was standardization—the creation of written processes that allow the organization to not only sustain the gains made through the kaizen process, but also share them with other divisions."

I've had several university students of mine deliver presentations on 5S implementations at both Burlington Coat Factory and Wegmans grocery stores. The results in both cases were impressive in terms of organization, safety, and cleanliness.

Another example of 5S at Wegmans grocery stores was the company's reorganization of the frozen goods warehouse at its central bakery. The project reorganized the storage methods utilized in the freezer, implemented a prioritized storage system, and created a highly visual communication and information system [Burns and Hogan, 2011].

Managing Employees

The management of employees can make or break any business. One of the key philosophies of Lean is that everyone needs to participate in continuous improvement in order to methodically identify and eliminate waste.

Therefore, it is critical from a variety of perspectives to utilize personnel properly, including:

1. *Labor costs*—Wages and benefits may account for up to 50 percent of operating costs in retail.
2. *Employee turnover*—This can be very high in retail, often resulting in increased recruitment, training, and supervision costs.
3. *Poor recruitment and hiring practices*—These can result in suboptimal sales skills, poor treatment of shoppers, and data errors during sales and inventory transactions
4. *Improper utilization of technology*—There have been significant advancements in technology that can greatly improve productivity, yet some retailers are still very labor-intensive.
5. *Poor labor scheduling*—This can result in a retailer's being over- or understaffed.
6. *Union restrictions*—There may be less flexibility for firms with unionized employees.

[Berman and Evans, 2013]

So what can we do to avoid these pitfalls in operations? According to Berman and Evans, it is important to consider and properly manage the following areas:

- *Hiring processes*—Doing a better job of recruiting and hiring can reduce turnover and result in better performance.

- *Workload forecasts*—Employee scheduling is one of the key ways for a service industry to ensure proper cost control and service to customers. It is critical to be efficient and effective when doing this. There is plenty of scheduling software available to optimize this process and minimize waste (e.g., www.kronos.com).
- *Job standardization and cross-training*—As previously mentioned, standardization is a key Lean tool to minimize variability, and cross-training helps to both balance the workload and minimize labor costs. In retail, there are many opportunities for both, especially in similar jobs such as cashiers in different departments. Cross-training (e.g., as a cashier and a stock person) adds flexibility, reduces the number of employees needed, improves motivation, and reduces boredom.
- *Good communications*—It's hard for employees to do their jobs well if they don't understand what is expected of them in terms of company policy and don't feel connected to the company as part of a team.
- *Employee performance standards*—Again, it's hard for employees to do their jobs well if there are no standards of performance. Personnel should be measured using standards such as checkout speed, data input, markdowns, and even sales goals and profits. When someone knows the specifics of what's expected of him or her, that person tends to be more productive and is usually happy working to reach those goals (if the person is given the tools and training to reach the goals and if they are reasonable, of course).
- *Compensation*—Whenever I do Lean training, one of the first questions I'm asked by employees is, "What's in it for me?" As we all know, money, promotions, and recognition that reward good performance (even a pat on the back) can help to motivate employees. For example, if a cashier is paid a bonus based upon reducing errors at checkout, that person will be highly motivated to reduce errors.
- *Self-service*—Operating costs can be reduced by automation such as self-service at checkout (e.g., one employee may be able to monitor four self-checkout lanes but operate only one cash register). However, we must be careful when considering and implementing this, as customers may feel that there is inadequate customer service or be frustrated by machine "glitches" (e.g., the scanner can't read the bar code). The retailer may also lose some opportunities for cross-selling and upselling of other complementary items.
- *Length of employment*—In general, increased length of employment is a good thing, as longer-term workers are more knowledgeable and productive (although they are usually more highly paid). If this is not managed properly, however, we can face the waste of underutilized employees, which can develop over time through improper support, training, and communication with the employees.

One example of a company in the service industry that does a great job of focusing on hiring and managing employees effectively to get the greatest performance out of them is Hard Rock Cafe restaurants. Its value system is designed to bring a fun, healthy, nurturing environment into the Hard Rock Cafe culture.

The company's strategic plan includes building a culture that allows for acceptance of substantial diversity and individuality and promotion from within (in fact, 60 percent of managers are promoted from within). The company's award-winning training is very detailed, with interactive DVDs covering kitchen, retail, and front of the store service.

Applicants are screened based upon their interest in music and their ability to tell a story. The goal is to hire bright, self-motivated people with an employee bill of rights and employee empowerment. In the end, this contributes greatly to the firm's low employee turnover rate, which is 50 percent lower than the industry average [Heizer and Render, 2013].

Managing Merchandise

In this section, we will discuss store design, layout and visual merchandising, and security and maintenance, all of which have a significant effect on the overall efficiency and effectiveness of a retailer.

Inventory

Although the role of inventory management in merchandising was covered in Chap. 8 (more from a planning perspective), there are still operational issues to consider, such as:

- How can you best coordinate deliveries to the store, especially from multiple suppliers? Smaller, more frequent deliveries allow for more of a just-in-time (JIT) inventory approach at the store level.
- How much inventory should you keep on the shelves versus in the back storeroom (or even warehouse)? Most stores have minimal back storeroom space. Thus, the JIT approach just mentioned helps to maximize revenue and service levels. There is even a trend for retailers to use radio-frequency identification (RFID) to help not only manage warehouse inventory, but also better manage the flow from the back of the store to the front. In fact, "Global clothing retailer and manufacturer American Apparel reports that it intends to equip all of its 280 stores with RFID technology. The system uses passive RFID tags to track every item as it enters the store, is moved to the sales floor, and then proceeds to the (POS); the system has proven to increase inventory accuracy and reduce the incidence of shrinkage due to employee theft or error" [Swedberg, 2012].
- What inventory functions, such as cycle counting, can be done after hours? This addresses worker productivity, as there are certain activities, like cycle counting that can be done during less busy time periods.
- How much support will you get from suppliers in setting up merchandise on shelves and displays? Most manufacturers will offer at least some support, including planograms (a visual representation of a store's products or services) and on-site assistance in setting up shelves and displays for their products.

The answers to these types of questions will affect the other areas of store operations that we will be discussing, such as design, layout, and visual merchandising.

Store Maintenance and Housekeeping

Store maintenance refers to the upkeep and repair of the shop and includes all activities to manage physical facilities. Housekeeping refers to the general cleaning of the shop and is carried out daily.

Store Maintenance The quality of store maintenance affects consumer impressions, the life of the facility, and operating costs. Preventive maintenance can extend the life of facilities and equipment for a longer period before having to invest in new ones and/or engaging in major overhauls.

It is important to consider who has responsibility for maintenance and repair of the facility (both inside and out) and the equipment, as this can have a significant impact on operating costs. If it is your responsibility, you can choose to do this with your own staff or to outsource the service (or some of it may be part of a lease agreement).

If you are outsourcing maintenance and repair, it's a good idea to look at the total cost of ownership of equipment (i.e., direct and indirect costs of a product) and to analyze the trade-off between paying for a monthly service agreement and the cost of having more breakdowns because of a lack of adequate maintenance.

Energy costs are also an area that can produce "Lean and Green" results. Many retailers and distribution centers are going with solar panels to reduce electric costs.

Also, as lighting accounts for much of a retailer's energy costs (up to 90 percent in some cases), more efficient lighting can pay off in two ways: (1) through lower energy costs using LED bulbs, and (2) through lower air conditioning costs, since new bulbs give off less heat.

Store Housekeeping Housekeeping is carried out at a number of critical locations and can be divided into front-end and back-end tasks.

Front-end tasks include:

- *Front end of store*—Keep areas just outside and inside the entrance of the shop clean.
- *Doors*—Keep doors clean and clear of unnecessary materials.
- *Signage*—Keep it clean, correct, and up to date.
- *Display windows*—Like doors, keep them clean, clear, and up to date.
- *Aisles and floor*—Keep them clear of clutter and mop or vacuum them daily.
- *Checkout counters*—Keep them neat and presentable, with no clutter. There can be minimal displays of add-on and promotional products.
- *Walls*—Keep them clean and free of cracks, discoloration, and the like.
- *Ceiling*—Check for water seepage or leaking. Correct and current information must be used as hang tags, aisle markers, and the like.
- *Lights*—Maintain them for proper functionality.
- *Air conditioning and heating*—Maintain the temperature at a proper, comfortable level.
- *Cash register and POS system*—Keep them clean and updated with current, accurate information.
- *Sound system*—Check for proper sound level and quality.
- *Fixtures*—Clean them regularly and make sure that they are in proper functioning condition.
- *Cleaning and maintenance tools*—Make sure that they are always available and properly marked, and that they have a "home" location.
- *Equipment*—Properly clean and maintain items such as mirrors, closed circuit televisions, and the like.

Back-end housekeeping tasks include:

- *Use of space for stock room*—This must be kept clear of clutter to maximize efficiency in the handling, checking, unpacking, and storage of goods.
- *Stock arrangement*—Stock must be organized with proper rotation and high-volume items closer to the sales area.
- *Food storage*—Food items must be properly stored and labeled and kept at appropriate temperatures.
- *Security*—There must be adequate lighting, and high-value items must be properly secured (i.e., locked, in a cage, or secured in some other manner).
- *Cleaning materials and equipment*—These must be stored and labeled away from merchandise (especially food).
- *Refuse area*—Refuse must be kept in sealed containers and disposed of daily.

- *Restrooms*—These must be continuously inspected and cleaned (especially when the retailer sells food products).

  [Tang and Lim, 2008].

Housekeeping is an ideal application for 5S workplace organization, which can ensure not only that the above-mentioned activities will get performed, but that "ideal" conditions will be maintained through tools such as end-of-shift/day cleanup (sort out), inventory check (set in order), and housekeeping and cleaning checklists (shine).

Merchandise Security

Inventory shrinkage occurs when the physical count of inventory in a store or distribution center is less than the perpetual or running (i.e., system) count. Control of inventories in retail and wholesale can be a critical component of profitability. Shrinkage can be caused by:

- Error—Through miscounting or computer input errors

- Theft—By customers, employees, or even vendors

We've already discussed the use of technology, such as bar code readers and electronic data interchange (EDI), and standardized work to reduce the waste of errors, but it is also important to have a good cycle-counting program to ensure inventory accuracy.

In cycle counting, items are counted and records updated on a periodic basis, often in conjunction with ABC analysis to determine the cycle. In general, A items are cycled through more often, as there are fewer of them, and they contribute more to revenue or profitability. B items are counted less often, and C items (the majority) even less frequently.

Cycle counting has several advantages over the more traditional practice of taking annual physical inventories, as it:

- Eliminates shutdowns and interruptions.
- Eliminates annual inventory adjustments.
- Has personnel auditing inventory accuracy.
- Allows the causes of errors to be identified and corrected.
- Maintains accurate inventory records.

In terms of theft, shrinkage prevention can be accomplished through a combination of education of employees, planning of the store or distribution center layout to prevent theft (especially of incoming and outgoing product), and the use of security personnel, devices, and preemployment testing.

The effect of merchandise shrinkage on the bottom line can't be overemphasized, as a 2011 National Retail Security Survey suggested that retailers in 2011 lost $34.5 billion to retail theft or shrinkage (i.e., the loss of inventory because of employee theft), shoplifting, paperwork errors, or supplier fraud. Overall, that accounts for approximately 1.41 percent of retailers' sales that year [Grannis, 2012].

Store Design, Layout, and Visual Merchandising

Store Design

According to Levy and Weitz, store design implements the retailer's strategy by meeting the needs of the target market, building a sustainable competitive advantage, and displaying the store's image.

They list the following objectives that are necessary to implement a retailer's strategy using store design:

- Build loyalty—This is done by meeting both customers' utilitarian needs (i.e., enabling customers to locate and purchase products in an efficient and timely manner with minimum hassle) and their hedonistic needs (i.e., offering customers an entertaining and enjoyable shopping experience).
- Increase sales on visits—Store design has a large effect on which products customers buy, how long they stay in the store, and how much they spend during a visit.
- Control cost—It is necessary to control the cost of implementing the store design and maintaining the store's appearance. As a result, store design influences the shopping experience and thus sales, labor costs, and inventory shrinkage.
- Legal considerations—It is necessary to protect people with disabilities from discrimination in employment, transportation, public accommodations, telecommunications, and requirements of state and local government.
- Design trade-offs—There is a trade-off among ease of locating merchandise for planned purchases, giving customers adequate space to shop by exploring the store and making impulse purchases, and the productivity of using limited space for merchandise.

Efficient and effective store design can have a huge influence on both maximizing revenue and minimizing costs by creating a retail image that the customer enjoys.

As we're talking about continuous improvement in this book, Lean thinking can be of great help both in the initial store design and on an ongoing basis through store layout and visual merchandising to reach the retailer's overall and design objectives, which we will now discuss.

Store Layout

The layout of a store basically consists of the arrangement and placement of fixtures, equipment, merchandise, aisles, and non-selling areas such as checkout, fitting rooms, and the receiving/stockroom area.

There are a few basic factors to be considered when designing the store layout, both in general and also from a Lean perspective. When using Lean thinking in terms of layout, we need to ask ourselves, "Do we have what the customers want, where they want it, when they want it, and in the quantity they want it?"

We also need to think about our layout in terms of the five transition points in retail that were discussed in Chap. 1 (i.e., enter, seek, find, select, and transact), with a goal of reducing or eliminating wastes that can occur during these activities, such as excess motion, waiting, lack of information, defects or errors, overprocessing, and excess transportation.

Some of the early adopters of Lean retail, such as 7-Eleven and Tesco, have seen inventory turns, margins, comparable store sales and full-price sell-through increase considerably as a result of such initiatives.

Front-End (Sales Floor) Layout

There are three main types of general layout pattern that are appropriate for different categories of merchandise, each of which can have a major impact on sales and efficiency.

- *Grid (Fig. 9.3)*—This type of layout uses long displays to help guide customers through their purchases. Grid layouts make it easy to locate merchandise, allow more merchandise to be displayed, and are cost-efficient, but don't encourage customers to explore the store (i.e., there are limited sight lines to merchandise). This layout is suitable for grocery, discount, and drugstore retailers.
- *Free-flow (or free-form; Fig. 9.4)*—Fixtures and merchandise are arranged openly to encourage customer browsing. Fixtures and aisles are arranged somewhat unevenly. A free-flow layout provides an intimate, relaxing

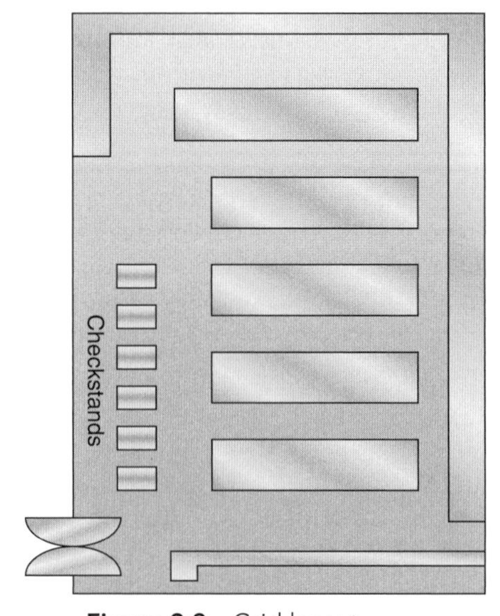

Figure 9.3 Grid layout.

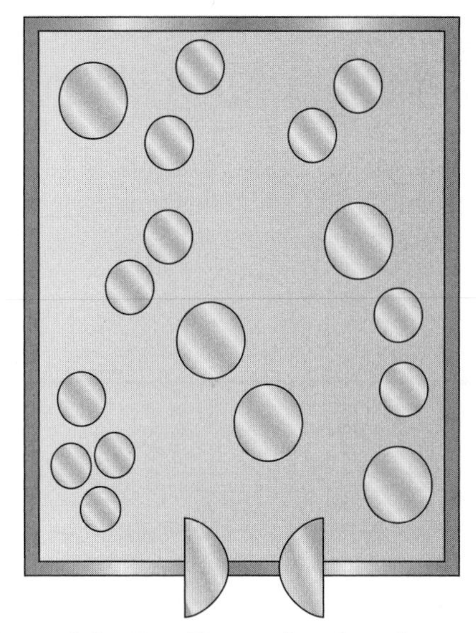

Figure 9.4 Free-Flow or free-form layout.

environment that facilitates shopping and browsing, but this pleasant, relaxing ambiance doesn't come cheap, as it may be an inefficient use of space and be more susceptible to shoplifting, as salespeople can't view adjacent spaces. It is typically used in specialty stores and upscale department stores (e.g., Reebok, Nike, and Disney stores).

A variation on this is the "boutique" layout. It is similar to free flow, but the departments or sections are arranged in the form of individual specialty shops, targeted at special market segments. This is not suitable for retailers with limited floor space.

▲ *Racetrack* (*or loop; Fig. 9.5*)—This layout has a loop with a major aisle that has access to departments and draws customers around the store. It provides different viewing angles and encourages exploration and impulse buying. This is typically used in department stores.

There are also some other layouts, such as the spine layout, where the major customer aisle runs the length of the store (e.g., Benetton) and the herringbone layout for narrow stores, where there is a single two-way aisle along the length of the store, with side roads leading to the walls from it (e.g., J&R Music World).

On a more "micro" level, we want to carefully develop and analyze a planogram, which, as mentioned earlier, is a diagram (sometimes computer-generated) that shows how and where specific stock keeping units (SKUs) should be placed on retail shelves or displays, within a product category, to increase customer purchases. This is especially critical when you are trying to draw your customers' attention to private-label (i.e., store) brands rather than national brands.

Figure 9.5 Racetrack or loop layout.

Back-End (or Back-Room) Layout

The back room contains the supporting materials and services that make the sales transaction successful. It contains the inventory of products that eventually will be moved to the sales floor, including supplies that are used in the retail process but not sold to the customer, and the things that managers and employees need in order to do their jobs. Managing the back room is an important task in retailing. It can be critical to profitability.

A retailer's goal is to reduce or eliminate back-room inventory that is not sold on the floor (e.g., expired or inactive items). That's because if inventory isn't out on the sales floor, it has much less chance of selling. A customer can't buy what he or she doesn't know about or see.

It is important to make sure that you have at least a few of every product that you sell out on the floor. If your business has high-velocity sales or slow restocking times, then you may require an extensive amount of redundant stock. However, keeping back-room stock also means that you must check the sales floor frequently and replenish it with stock from the back room.

The back room will also be used for the storage of supplies, such as cash register tape and shopping bags, as well as various specialty supplies needed by specific types of retailers. These supplies must be tracked and ordered just like finished goods inventory, so they must be able to be easily found when needed.

The typical back room will also house-support services requiring office space. This includes a place for employees to take a break or eat lunch as well as holiday signs, bathrooms, administrative offices, and the like.

All this requires room and takes away from the space available for sales, of course, but it is crucial to productivity and profitability, and it needs to be planned for. So it is important that the management and replenishment of materials in the back room is integrated with the front-end merchandise management tools discussed in Chap. 8.

This is another place that is ideal for a 5S workplace organization kaizen event, as the back-room area tends to get cluttered and dirty easily.

While store layout attempts to maximize profit through product exposure, there is something known as a "servicescape" that considers the totality of the ambience and physical environment in which a service occurs.

A servicescape considers ambient conditions such as (1) lighting, sound, smell, and temperature, (2) spatial layout and functionality, involving customer circulation path planning, aisle characteristics, and product grouping, and (3) signs, symbols, and artifacts that may have social significance to customers.

These factors are all considered under the concept of *visual merchandising.*

Visual Merchandising

Visual merchandising is the development of floor plans and three-dimensional displays and signage to maximize sales. Goods or services can be displayed to emphasize their features and benefits. Visual merchandising is used to attract, engage, and motivate the customer to purchase a product or service. This includes the selection and placement of signage, fixtures, counters, checkout, fitting rooms, displays, end caps, and promotional aisles.

Signage

Signage acts as a virtual salesperson and can help to improve the shopping experience (e.g., give directions to customers), minimize transportation and motion waste on the part of the customer, and enhance sales. Signage can be found in a variety of places, such as service and checkout counters, fitting rooms, restrooms, and product departments, and it performs the functions of:

- *Location*—Identifies the location of merchandise and helps to direct customers.
- *Category*—Identifies types of products. This type of signage is usually located near the product.
- *Promotional*—Usually relates to specific offers and may also be found in windows.
- *POS*—Is near sales merchandise, and usually gives prices and product information.
- *"Lifestyle" images*—Tries to create a feeling that encourages customers to shop.

A method for further reducing your signage costs and at the same time being more appealing is the use of visual content delivered digitally through a centrally managed and controlled network and displayed on a TV monitor or flat panel screen. This can help to attract attention and enhance the store's environment by providing an appealing atmosphere. The message can target specific demographics and can eliminate the costs of printing, distributing, and installing traditional signage.

Fixtures

It is important to use appropriate fixtures and fittings that enhance the value of the merchandise and facilitate ease of maintenance.

Some general types of fixtures are:

Straight rack—These hold a lot of apparel, but it's hard to feature specific styles and colors when using them. They are usually found in discount and off-price stores.

Rounder (bulk fixture or capacity fixture)—These are smaller than a straight rack and hold a fairly large amount of merchandise. They are easy to move around, but customers can't get a front view of the merchandise.

Four-way fixture (feature fixture)—These hold a relatively large amount of merchandise and allow customers to view the entire item. They are hard to maintain because of the variety of styles and colors and are used primarily by fashion-oriented apparel retailers.

Gondolas—These are versatile and are used primarily in grocery and discount stores (e.g., to hold fresh vegetables in a supermarket). It can be hard to see apparel, as it is usually folded.

Being innovative and creative with fixtures can reduce inventory and improve productivity.

Many retailers are focusing on efficient product packaging and in-store fixtures to reduce stores' handling costs. Fixtures that use gravity to feed products forward, for example, can cut detailing time for employees in the store.

In grocery retail, the supermarket is the largest opportunity for savings, as that's where the most of the labor costs occur. Industrywide, store labor is 11.9 percent of store sales.

Pathmark Stores, in Carteret, New Jersey, is always on the lookout for ways to reduce store handling to cut labor costs. It has a team focused on store layouts and fixtures to improve customer flow and product location.

Retailers are looking at the increased use of gravity feed fixtures in the soft drinks area, which can be worth $30,000 to $40,000 per year in savings for an average store in some cases. For example, Save Mart Supermarkets of Modesto, California, is testing gravity-feed fixtures for refrigerated cases in the dairy department and is also testing rack systems that roll the rack into the case. The firm thinks these efforts could reduce store handling costs by 3 to 5 percent annually.

Another area that retailers are investigating to reduce handling costs is filling products from the back of the case.

Finally, tilted shelving can help cut stores' handling and reduce inventory costs substantially (and, as an added bonus, displays the packages well) [Purpura, 1998].

Customer Flow

The longer a customer is in a store, the more likely he or she is to buy something. So, the aisles should be wide enough to encourage comfortable movement.

When considering layout, we need to keep in mind that the objective is to maximize profitability per square foot of floor space and that sales and profitability vary directly with customer exposure.

So, for example, in a supermarket, it is typical to locate high-draw items around the periphery of the store, using prominent locations for high-impulse and high-margin items. Supermarkets also tend to distribute power items to both sides of an aisle and disperse them to increase viewing of other items, and to use end-aisle locations. Some even like to convey the store's mission through careful positioning of a lead-off department such as the bakery.

There is also the concept of experiential shopping, where the shopping experience is made more interactive by using tools such as demonstrations, workshops, and samples. Retailers such as Home Depot, Williams-Sonoma, and major supermarkets use this tool extensively.

Cashier/Counter

The physical location of a counter depends on both the size and shape of the store and customer convenience.

In a smaller store, the cashier is usually placed close to the entrance, whereas in a large department store, the counter may be more in the center of the store, where it can also serve as a customer service counter.

In most supermarkets, the checkout counter is located near the end of the shopping trip to help with the payment process.

A narrow store tends to have the counter against the wall to increase the space available to customers and for product displays.

Fitting Rooms

It may surprise some people that the fitting room is an important factor in the customer's decision to make a purchase. What's not a surprise is that many retailers spend more time and money on their store interiors and often neglect the changing room.

There are many improvements that can be used to enhance the fitting room:

- *Lighting*—Instead of regular lightbulbs, try normal day lights and disco night lights to facilitate the customers' buying decision.
- *Space planning*—Make sure that the fitting rooms are big enough for customers to change in comfort and view their clothes properly.
- *Mirrors*—They should be clean and not cracked or distorted.
- *Use of 5S*—Fitting rooms should not be cluttered with boxes or have broken hooks or torn carpets, which can cause accidents as well as give the customer a negative impression. A clean, safe fitting room can add value to the product. So this is another great area for a 5S kaizen event.

Window Display

The front window is a place that potential customers pass constantly and should be thought of as advertising space that needs to be updated on a regular basis. In the case of a smaller location, a "no-back" window can add light to the store, and as a result, the interior becomes the "window display" for the shop.

Lighting, Sound, Smell, and Color

A retail environment includes things that appeal to the five senses and contribute to the brand image and surroundings of a retail store. So the presentation of products for selling in an atmosphere that triggers the shopper's desire to buy is very important.

Shoppers sense the environment when they enter the store. From the front entrance to the back wall, the environment should communicate the store's identity or brand image. Everything in the store—the signage and graphics, the retail store fixtures, and the merchandise displays—should work together cohesively to tell the customer about the store brand.

Examples such as McDonald's restaurants and Bath & Body Works infuse the store atmosphere (literally) in a positive way to influence the buying behavior of their customers.

We'll talk more later about getting started with Lean, but in general, it's always a good idea to:

- Start small and pick a single pilot area or department (e.g., cosmetics or pharmacy).
- Develop a better understanding of the consumer experience and the shopping process, including flow, aisle penetration, dwell time, product interaction, and queue lengths.
- Identify waste.
- Use a team-based approach to get ideas from the people who are on the floor and have great ideas.
- Expand beyond the pilot to other areas in the store.

Customer Service

Customer service is the activities and programs that allow consumers to receive what they need or desire from your retail business. It is important for a sales associate to greet the customer and make himself or herself available to help the customer find whatever he or she needs. When customers enter the store, it is important that the sales associate make customers feel welcome and make sure that they leave the store satisfied. For retail store management, training for yourself and your staff enables you to have excellent customer service skills.

Good service encourages customers to return and can generate positive word-of-mouth communication, possibly attracting new customers.

At the end of the day, providing consistent high-quality service gives retailers an opportunity to develop a sustainable competitive advantage.

Customer Service Program

In Chap. 7, we discussed having a customer relationship management strategy. A direct operational offshoot of that is your customer service program. The program should include a clearly defined set of service objectives that guides the retailer's efforts on service features and actions. The types of services provided and the quality of service delivery should also be defined.

Retailers usually choose between a standardized and a personalized approach to customer service. Higher-end stores such as Nordstrom's tailor service to meet each customer's personal needs. This can be more costly and also less consistent. General retailers like Walmart or McDonald's offer a more standardized approach to customer service that is based upon establishing rules and procedures and then making sure that they are implemented consistently.

In either case, it is necessary to clarify your service standards to your employees, to provide a guide for decision making and build a service-oriented culture at the company. This also requires a team of company leaders and employees who are all "on the same page." The service strategy should help to differentiate the company from its competitors and requires constant support and training. A good place to start is with a service audit to determine what is important to customers.

As part of the program, visible service symbols, service incentives, and measurements must be explained and communicated to all levels of employees, along with constant training on customer service and problem resolution skills.

Customer Service Lean Training

An interview with Devesh Raj, a partner and managing director at the Boston Consulting Group (BCG), pointed out:

> [A good] example of Lean customer service in retail is a customer support call to retail banks. This process has cyclicality in work volume because calls to retail banks will often go up in volume and come down. Often these peaks are predictable. So in the retail bank that we worked with, for example, the middle of the week is when the calls peak. If you don't handle that peak time with extra capacity, you can lengthen the average wait time of the call, the customers will get frustrated and they can drop off. That's not a good outcome. Companies will generally handle this with additional staffing around peak times, but those people are not doing much the rest of the days. A simple idea that you would see in lean is the idea of cross-training across different groups to smooth out the

utilization. So for example, you could take the mortgage desk, cross train them on handling support calls and bring them into service whenever the call volumes peak. Then they can go back to their work on the other days. This is a simple example that we've seen work in the banking space and in most customer-service operations.

These examples of cross-training and tackling uneven processes grew up in the manufacturing world, but can be very powerful in a services context. [Raj, 2013]

Identifying and Closing the Customer Service Gap

Levy and Weitz suggest using a gap model to improve retail customer service quality. It recommends looking for gaps in terms of:

- *Knowledge gaps*—Find out what the customer wants through a variety of means, such as surveys, customer complaints, feedback from employees, and so on. This information should be used to improve customer service.
- *Standards gaps*—Define the role of service providers, then set and measure new targeted service goals. This requires extensive information sharing and training.
- *Delivery gaps*—Meet and exceed service goals through employee training and incentives.
- *Communications gaps*—Communicate the service promise to the customers and manage their expectations.

Empowerment

Besides being provided with training and tools (analytics and technological) to help encourage great customer service, workers need to be empowered, meaning that employees at the firm's front lines are allowed to make important decisions regarding how service is provided to customers.

To maximize the effectiveness of empowered individuals and teams, managers need to provide adequate support, including necessary training and clear objectives and goals, and, perhaps hardest of all, they must release some control.

There is also the question of motivation, as most people want to know, "What's in it for me?" To answer this question satisfactorily, businesses can use a variety of financial and nonfinancial rewards, previously mentioned in Chap. 7, such as:

- *Bonuses*—Cash or stock options
- *Profit sharing*—Profits distributed to employees
- *Gainsharing*—Rewards for improvements
- *Incentive plans*—Typically based on production rates
- *Knowledge-based systems*—Rewards for knowledge or skills

Sometimes just a pat on the back, a celebratory meal, or an employee of the month award can be enough to make the employee(s) feel appreciated.

Lean Customer Service Examples

An example of flexibility in managing the waste of waiting found in a checkout line is the fact that many retailers with multiple checkout lanes in their stores set standards to reduce the time that customers have to wait in line. Many customers will leave a store and not patronize the retailer in the future if they feel that the wait to checkout or talk to a sales associate is too long or unjust.

In reality, research has found that customers overestimate the time they wait in line by as much as 50 percent and that the estimates are affected by store atmospherics such as a pleasant aroma.

The fact is that the fastest way to process customers is to create a single line of customers and have the person in front go to the next open cash register—the system used at airports and occasionally in fast-food restaurants and banks.

The exception to this is at supermarkets, where to reduce wait time, stores often establish express lanes for shoppers with few items so that they don't have to wait behind customers who are checking out with large baskets of items. However, this practice has the somewhat odd effect of treating the best/biggest customers the worst [Piyush, 2005].

One customer service analysis at a retail pharmacy checkout found that the average time for a problem-free transaction was 46 seconds, but in this store's case, 80 percent of transactions had problems that led to longer transaction times.

As a result, delays and non-value-added tasks nearly doubled the transaction time to 1.4 minutes. In fact, 1 out of every 10 customers, on average, experienced one or more "time bomb" issues that required management intervention or significant problem investigation. These transactions averaged more than 2.5 minutes and had a negative impact on the customers waiting in the queue as well.

Overall, 70 percent of observed transactions experienced a delay caused by customer actions or process issues such as the following:

- The crew member waits to begin the transaction until the customer provides the loyalty card.
- The customer searches for the loyalty card.
- The crew member must look up the loyalty card using the customer's phone number.
- Items cannot all be staged at the same time.
- The customer searches for the payment method or exact change.
- There are cashier or system delays in processing a credit or debit card.
- The customer inspects the receipt or gathers belongings, delaying the next customer.

About 10 percent of observed transactions experienced a time bomb that negatively affected the speed of checkout, such as:

- An issue or confusion with loyalty offers and/or manufacturer's coupons.
- A customer who was returning or exchanging merchandise.

- A customer who couldn't find the loyalty card (and for whom phone lookup is not successful).
- A customer inquires about the price of an item, requiring validation.
- A call to management for assistance (issue resolution) or authorization.
- A customer leaves in midtransaction to gather an additional item.
- Management is called for change.
- A customer is issued a rain check after substitution offers are not accepted.
- It is necessary to address a declined check payment.

[Guidon Performance Solutions, 2009]

By collecting and analyzing checkout data, this retailer was able to identify a host of issues that contributed to 10 percent of its transactions having major issues at checkout. The next major step would be to do a Pareto analysis on the documented time bombs and come up with creative solutions to eliminate the major ones (i.e., do root cause analysis).

Before discussing how to get started on your company's Lean journey, we will examine opportunities for Lean in the warehouse, which both retailers and wholesalers can benefit from.

CHAPTER 10

Being Lean: Retail and Wholesale Distribution

As detailed in Chap. 2, the definition of a retailer is an organization that purchases goods or products in large quantities, either directly from manufacturers or through a wholesaler, and then sells smaller quantities to consumers for a profit, while a wholesaler is a business that buys large quantities of goods from various producers or vendors, stores them, and resells them to retailers. Wholesalers that carry only noncompeting goods or lines are called distributors.

Wholesalers and larger retailers have one thing in common: they both operate distribution centers (DCs, also known as warehouses), an area that is ripe for the implementation of Lean concepts.

After all, the whole idea of a warehouse is to add value to the customer by having what he or she wants at the right place and the right time. If anything gets in the way of doing this, waste is created. So it is important to be able to deliver quick, accurate order fulfillment at a low cost.

Lean thinking is relatively new and is applied less frequently in distribution. In fact, according to the Manufacturing and Wholesale Distribution National Survey [McGladrey, 2010], "distributors implement these [Lean] systems far less frequently than manufacturers (45 percent vs. 73 percent), Lean principles are applicable to any industry. With that in mind, distributors may wish to take a closer look at how lean can help their businesses become more efficient."

As Lean is still in its early stages in supply chain and logistics, it is sometimes difficult to find a place to start. One place that many companies have found is a good place to begin is in the warehouse.

In the 21st century, the warehouse is becoming a strategic tool to be used for competitive advantage, as opposed to just a place for long-term storage. Warehouses today are more often actually DCs supporting a just-in-time (JIT) supply chain that is low-cost, flexible, and efficient, especially in the rapidly growing world of e-commerce. E-commerce growth affects both the warehouse and the inbound and outbound logistics that support the facility [Myerson, 2012].

Eight Wastes Applied to Warehouse Operations

In Chap. 8, we discussed how an efficient, collaborative, technology-enabled forecasting and inventory planning process can reduce waste in both retail and wholesale. The same goes for Lean warehouse operations. Let's start by considering sources of waste from a warehousing perspective:

1. *Overproduction*—Performing warehouse functions too soon or before an order has been placed (and there may already be excess inventory in a warehouse because of overproduction or overpurchasing).
2. *Waiting*—As we know, waiting is a waste of time. This is as true in warehouse operations as it is in production. It can be applied to machines, materials, information, suppliers, customers, employees, and anything else that has the ability to sit and wait to be processed. As the saying goes, "Time is money," so a Lean operation should always try to reduce lead time.
3. *Transportation*—Unnecessary carrying and movement of material from one place to another is a major source of waste in materials handling.
4. *Overprocessing*—This includes having too much complexity or too many steps in your processes as well as overchecking. Many warehouse processes are riddled with extra steps, checks, and confusion.
5. *Inventory*—Excessive inventory is the result of overproduction and not only ties up the cash the firm has to run the business from, but also ties up its resources with the ineffective management of the surplus inventory. Also, poor inventory control (i.e., inventory inaccuracies) represents waste, resulting in frequent stockouts and late deliveries.
6. *Motion*—This occurs when work isn't planned and lined up sequentially to minimize the non-value-added motion that frequently occurs in a warehouse. Excess movement can also result in a search for tools or stored items that cannot be quickly and easily located.
7. *Defects or errors*—In warehousing, errors result in waste. A warehouse is a large process in the midst of an even larger process. Material can arrive packaged incorrectly or be of poor quality. Inventory counts may be off significantly. Customers may be shipped a poor-quality or incorrect item. To reduce the waste of rework and defects, a warehouse should build quality and standardization into its processes to prevent the passing on of errors.
8. *Behavior*—A frequently overlooked waste in warehousing is the waste of our employees' knowledge. If they have some warehouse operations experience as well as training in Lean, they can help to identify waste in their daily activity and be able to provide valuable ideas for improvement.

Lean Thinking in the Warehouse

As mentioned in Chap. 7, in "The Skinny on Lean," Peter Bradley cites a five-step process as a guide to implementing Lean principles, which can be applied to the distribution environment as well as to manufacturing.

The steps are:

1. Identify what your customers expect and determine what value you add to the process.
2. Plot the value stream. Identify all the steps involved in moving goods through the system.
3. Make the process flow.
4. Pull from the customer.
5. Pursue perfection. Root out any remaining waste [Bradley, 2006].

The fact is that most Lean concepts can work well in the warehouse, especially 5S (which is usually the first activity engaged in as a good foundation for a Lean program), value stream mapping (VSM), team building, kaizen, problem solving and error-proofing, kanbans/pull systems, line balancing, cellular applications, and general waste reduction.

At first glance, many warehouses are very neat and organized, at least in the case of "pure" distribution companies, although perhaps less so at manufacturers (it's not their area of expertise). However, once you "get under the hood," there are plenty of opportunities to be found.

"Assembling" Orders

Warehouse operations seem to be very active, with people and equipment in constant motion. However, this does not mean that they are productive.

Orders do not necessarily keep moving. They tend to pile up and sit waiting between processing steps, causing clutter and taking up space (all forms of waste!). An analysis of a distribution operation done around 2006, as described in "Are Your Warehouse Operations Lean?" by Ken Gaunt, showed that a typical order was being worked on for only 38 percent of its cycle time; 56 percent of the time, orders were idle; while the remaining 6 percent involved employees dealing with problems such as waiting for equipment, computer issues, interruptions, and blocked aisles.

To radically improve this type of performance, warehouse orders should be looked at as being "assembled" in the most efficient manner, minimizing non-value-added activities, including delays in receiving, putting away, picking, packing, and loading, as well as poor picking paths, wasted motion, congestion, and poor equipment condition and availability.

For example, the orders could be assigned based on the amount of time that it takes to pick batches of line items (established through time-and-motion studies), instead of just giving an entire order to a picker. Pickers would be assigned zones. They would then feed workstations at regular intervals to keep products flowing, so that packers and loaders are not kept waiting.

In general, you want to look at aisle and rack layout to improve space utilization, making sure that products are arranged so that the most frequently used items are closest to shipping to reduce travel distance; heavily use visual systems for aisles, racks, products, and workflow; avoid cluttered and blocked aisles; and create housekeeping systems to improve efficiency by ensuring that tools and equipment are available when and where they're needed [Gaunt, 2006].

Reverse Logistics

Reverse logistics is the process of moving goods from their typical final destination for the purpose of capturing value or for proper disposal.

In many cases, returns are a result of errors in shipments or product defects, a major form of waste in terms of Lean thinking. Reverse logistics can also be the result of a product needing servicing or a customer just not wanting the item, as well as a whole host of other reasons such as processing returned merchandise that is damaged, dealing with seasonal inventory, restocking, salvage, recalls, excess inventory, recycling programs, hazardous material programs, obsolete equipment disposition, and asset recovery.

No matter what the reason, reverse logistics exists in wholesale and retail distribution (and retail store operations) and must be managed as efficiently as possible.

According to Karen Hawks of Navesink Logistics [Hawks, 2006], retailers are now using return policies as a competitive weapon (from a shopper's perspective, it's always a good feeling to know that the retailer has liberal return policies).

Grocery retailers were an early pioneer in the area of returns, as their profit margins are slim, so they require a good return management system. They created reclamation centers that led to centralized return centers (now fairly common). Because of consolidation, some large retailers today are more powerful than manufacturers (e.g., Walmart), and are therefore able to force the liberal return policies that they offer to customers back onto manufacturers.

To improve the efficiency of the returns process, retailers tend to use bar codes, computerized return tracking and entry, electronic data interchange (EDI), and radio-frequency (RF) technology in their reverse logistics management more frequently than manufacturers do.

It's not hard to see why this is such an important process to consider, as returns can reduce retailers' profitability by more than 4 percent.

The extent of reverse logistics functions varies based on industry and channel position (and whether returns are a larger percent of operating costs).

An example where it is hugely important is the publishing industry, where returns can be the difference between making and losing money. Publishing has the highest rate of unsold copies (28 percent on average). Factors contributing to this are both the relatively short shelf of publications and the fact that superstores sell less than 70 percent of the books that they order.

Another example of an area where the returns process is critical is technology retailers like Best Buy that sell products with a short product life cycle, and therefore also need a fast return process. They also face the challenge of recycling and in some cases reusing or remanufacturing old equipment (e.g., ink cartridges).

Some thoughts on elements to look at in terms of reducing the cost of reverse logistics would include:

- *Avoidance*—Design merchandise and systems in a manner that will minimize returns to prevent customers from sending purchased products back. Tactics would include improved quality, establishing return agreements with trading partners, and improved customer service, such as providing toll-free numbers that customers can call before returning products.
- *Negotiation*—This is an important element for all parties involved in the reverse logistics process. There is a general lack of expertise on product returns, so negotiations are usually approached without formal pricing guidelines. Organizations often do not maximize the residual value of returned products.
- *Financial management*—Returns are sometimes charged against sales, so the sales department may tend to fight returns and delay them as much as possible. Accounts receivables are also affected by returns.
- *Outsourcing*—Reverse logistics is usually not a core competency of the firm. Thus, it may make more sense for the company to outsource its reverse logistics function than for it to keep the function in-house.
- *Gatekeeping technology*—Include technology for screening for defective and unwarranted returned merchandise at the entry point into the reverse logistics process.
- *Disposition decisions*—Reduce the amount of time required to figure out what to do with returned products once they arrive.
- *Processing and cycle times*—Use Lean principles and tools to map the entire process and look for improvements throughout, including the return merchandise authorization (RMA) process, the reverse logistics network (i.e., central versus regional or local returns), and store or DC returns processing.
- *Data management*—The system should create a database at the store level so that the retailer can begin tracking returned product and follow it all the way back through the supply chain. It should also include detailed information about important reverse logistics measurements, such as returns rates, recovery rates, and returns inventory turnover.

Hawks suggests some other areas where you can look for improvements in your reverse logistics process as well:

- Streamline turn-in procedures.
- Route items with an eye to what happens to them next.
- Integrate the forward and reverse pipelines.

- Explore the potential of commercial software applications or techniques for improving reverse flow management.
- Align financial incentives with improvements.

[Hawks, 2006]

DCs (and stores) are traditionally built to ship things out, so in many cases, not much thought is given to the return, processing, repair, and replacement of products, which can be an area of considerable waste. If the reverse logistics process is operated efficiently, it can not only improve profitability but also add value to the customer.

Value Stream Mapping in the Warehouse

A good way to attain better flow in distribution is to start with VSM. The value stream map will give employees an overall view of all warehouse activities, allowing them to suggest improvements in other areas as well as their own. It is a good way for everyone to understand and agree on how the facility works and to come up with ideas for improvement. It's also a good idea to display the current- and future-state maps in the warehouse so that employees are able to see previous improvements and take part in the ongoing effort.

The current-state VSM shows the "as is" warehouse operation and will serve as the basis for the future-state changes. The map begins with the customer and works back through the warehouse process from shipping all the way back to the receiving area and even to key suppliers. To keep things simple, VSMs start with one product family or service.

A current-state map may identify opportunities to reduce lead time and processing time in the warehouse, such as:

- Decreased material handling time in order picking, putaway, and palletizing
- Issues with the strapping and metal detection machines
- Quicker truck loading time
- Less time spent checking inventory location and aging

The goal when developing the future-state map is to have continuous flow, reduce lead time, and eliminate as much waste as possible.

To develop a balanced continuous flow process by documenting where material and information don't flow properly, takt time (i.e., how frequently a unit must be completed to meet customer demand) for the demand stream is calculated and used.

A future-state map may identify improvements such as:

- A redesigned pick and pack area that reduces wasted motion and errors
- Labeling and zoning of warehouse racks to reduce time spent looking for products
- Bar coding of cases to track all cases and pallets in the warehouse
- Placing faster-moving products on lower rack levels and closer to the shipping dock (known as *velocity slotting*)
- Using RF bar code scanners instead of paper and manual data entry and implementing a warehouse management system (WMS; discussed later in this chapter) to obtain real-time inventory data and develop pick sequences to reduce order picking time and errors
- Increased employee cross-training to improve productivity and reduce errors

These types of Lean improvements, when implemented in a warehouse, can reduce order processing time by as much as 50 percent and lead time by up to 25 percent or more.

To successfully assess the operation using a value stream map, you need to:

- Involve the operators and supervisors.
- Identify Lean improvements and kaizens.
- Question every activity.
- Treat the warehouse like a large staging area.
- Develop justification as you go along.
- Implement the Lean improvements using the VSM plan.

Then start the cycle again.

Using 5S or workplace organization ("a place for everything and everything in its place"), while seemingly simple, is a good activity to start with, and it is sometimes surprising, once you document it, how much time people waste searching for material, tools, and information.

Lean Tools in the Warehouse

In order to get any of this done, a team approach is necessary to identify the wastes in areas such as errors (receiving, putaway, picking, loading, and so on), inventory inaccuracy, damage, safety issues, and lost time. Lean tools, such as problem solving (root cause analysis, fishbone diagrams, and so on) and error-proofing with standardized work (e.g., visual instructions on how to use a strapping machine or how to load or unload a truck), can also be helpful in the warehouse environment.

Pull systems using kanbans are a "natural" in a warehouse for everything from packing materials to forms, as well as product assembly and kitting.

If you do any value-added activities, such as kitting or assembly, as many third-party logistics organizations (3PLs) do, then the use of work cells might be appropriate to minimize labor and maximize the use of equipment and space. Line balancing is a tool that can be used in this type of situation, not only in staffing, but also to ensure proper flow in the work cell.

There's even a place in the warehouse for total productive maintenance (TPM, or reduction of equipment-related waste), as there's plenty of equipment (some of it automated) that might not be running as efficiently as possible (e.g., carousels, forklifts, hand trucks, strappers, and the like).

Lean Warehouse Examples

In recent years, retailers have been driving their networks of warehouses and distribution centers to become Leaner. Many have implemented a combination of technology, improved processes, and standards to trim inventories 20 percent in the past 15 years, according to Census Department data.

In most retail and wholesale warehouses, workers spend their days assembling shipments for individual retail outlets. Much of their time is spent traveling from place to place. Technology has been used to help reduce this travel time. In the past, workers would plan their own routes to retrieve items, using their knowledge and judgment to put orders together. Today, computers direct workers where to go at Sysco, the nation's largest restaurant supplier, and many other major retail and wholesale distribution centers.

Other methods, such as time-and-motion studies, are being used to generate standards and improve processes at places like a Rite-Aid DC in California [Brenner, 2010].

Amazon.com, one of the largest e-commerce businesses today, has used Lean methods in its distribution centers from the very beginning, as:

> It goes without saying that Amazon's warehouses are tremendously efficient to handle all of that activity without missing a beat, even in traditionally busy times for retailers such as Christmas.
>
> Much of Amazon's warehouse success is owed to the kaizen experts that run these operations. As the *Financial Times* reports, Amazon's warehouses process more than 35 orders per second, so operational efficiency is paramount to getting all these packages out the door. At each DC, a continuous improvement manager is assigned to apply Kaizen techniques to various processes and practices to streamline workflow.
>
> In Staffordshire, England, Matt Pedersen is the resident six sigma "black belt." He told the *Financial Times* that he walks his warehouse on a daily basis to identify potential wastes and opportunities to improve efficiencies.
>
> "We go to the associates and find out what's stopping them from performing today, how we can make their day better," Pedersen told the news source. Additionally, Pedersen deploys mobile problem solvers—employees that walk the warehouse with laptops—to assist in the process.

> Pedersen notes that Amazon expects a lot from employees—it's a global leader in the ecommerce field and expects its workers to keep up with operations. Ultimately, the use of kaizen helps employees keep up with the breakneck pace of 35 orders per second. [Enna.com, 2013]

Menlo Worldwide Logistics, a major 3PL provider, not only has implemented Lean at many of its facilities, but also uses it as a competitive weapon, as can be seen at its Web site under "Lean Logistics" (www.con-way.com), where it points out the following areas where it looks for waste:

- *Mapping material flows*—In studying material flows from raw material vendors to customer finished goods, we challenge each point at which material flow is stopped. Methods to speed up material flows include shipping ocean containers directly from Asia to inland regional warehouses, instead of transferring cargo to trucks at the ports of entry, and bypassing warehouses with large orders that travel directly from factory to customer.
- *Keeping drivers and tractors moving*—The interface between warehousing and transportation can often result in waste. We attack this problem by working with our carriers to drop trailers, freeing the driver and tractor to pick up a loaded one and keep moving. Through careful dock scheduling and synchronization of warehouse workflows, we can live-load shipments quickly, minimizing driver dwell time.
- *Using milk runs*—Milk runs reduce transportation costs and build more consistency into an inbound supply network.
- *EDI*—Menlo Worldwide Logistics makes extensive use of EDI and RosettaNet to pass data among our supply chain partners. Communicating data electronically eliminates errors caused by manual data entry.
- *Warehouse efficiency*—Menlo Worldwide Logistics' team of industrial engineers designs warehouse layouts that streamline inbound and outbound flows, maximize labor efficiency, and deliver high space utilization. We employ techniques like dynamic slotting, cluster picking, task management, and system-directed putaway to optimize labor and space efficiencies.
- *Optimize transportation routes*—Logistics Management System (LMS) application of Menlo Worldwide Logistics optimizes each load to meet delivery dates with low-cost mode and carrier selection.
- *Packaging optimization*—We work with customers to explore the use of returnable containers for repetitive shipments to factories. For finished goods, we can study packaging sizes to uncover ways to increase pallet and trailer utilization. Small changes in carton sizes can facilitate better storage utilization and lower transportation costs [Menlo Worldwide Logistics, 2011].

Furthermore, Menlo emphasizes the use of mistake-proofing tools such as:

- Making it easy to do things right and making it hard to do things wrong
- Easy-to-read visual controls

- RF devices coupled with bar-code technology
- System-directed cycle counting at our warehouses
- Utilization of Six Sigma and statistical process control (SPC)
- International Organization for Standardization (ISO) processes
- EDI
- Standardized processes
- Implement repeatable, standardized processes
- Establish one best way to perform each task
- Visual documentation of processes
- Correct any activity that causes rework, unnecessary adjustments or returns.
- Organized workplace (5S)

[Menlo Worldwide Logistics, 2011]

All of this results in benefits to Menlo's customers, such as better service, lower costs, higher availability, higher customer satisfaction, and more reliable deliveries.

This isn't just talk, either, as was pointed out by Gary Forger in "Menlo Gets Lean." The article describes how Menlo Worldwide Logistics operates a 250,000-square-foot facility in Michigan that recently shipped 8,000 orders in a two-week period with no errors and, according to a recent audit, has inventory accuracy of 99.99 percent.

This site was the pilot site for Menlo's Lean program (along with a dozen other Lean warehouses at the time), which had a goal of reducing the cycle time and increasing the productivity of various resources by eliminating waste.

Menlo focuses its metrics on service, quality, delivery, cost, and employee morale. Warehouse operators work in 20-minute segments or small "batches," similar to what was mentioned earlier in this chapter. That maximizes flexibility and allows labor to help minimize response time to orders. Items are slotted according to size and velocity, and workers are assigned certain aisles to keep neat and organized (and must sign off on a checklist). Team leaders "own" their processes, supervisors and managers remove "roadblocks," and the bonuses of hourly team members are tied to metrics and improved processes (a real key to success, I believe). Besides weekly departmental meetings to discuss performance and improvement, every month a kaizen event is held in which as many as six workers concentrate on improving an operation [Forger, 2005].

Peter Bradley, in "The Skinny on Lean," stated,

> Menlo Worldwide reports that warehouse productivity improved 32 percent between January and November last year, measured by gains in lines per hour. Defects, measured as the error rate, dropped by a whopping 44 percent. The on-time percentage for shipments was north of 99 percent in every one of those months, hitting 100 percent in eight of 11 months. And those involved think they can do more. [Bradley, 2006]

Another major 3PL player, Ryder Logistics, highlights its "Five LEAN Guiding Principles" on its Web site (www.ryder.com), which "provide the foundation for operation excellence, continuous improvement and supply chain efficiency." The guiding principles are people involvement, built-in quality, standardization, short lead time, and continuous improvement. Ryder also mentions using a variety of Lean tools in its business, such as workplace organization, visual management, work cells, standardized work, and even a Lean Academy. The company not only includes Lean applications in the warehouse operations, but starts by determining the optimal distribution network design, which can significantly reduce waste in the overall supply chain network. It takes this very seriously.

A Ryder case at the site describing how it took over Whirlpool's service operation showed how continuous improvement activities reduced costs, improved shipment accuracy and order cycle time, and boosted overall efficiency. The first thing the firm determined was that it was more efficient to consolidate Whirlpool's various service facilities into one location. After that, it implemented a variety of continuous improvement efforts, including the creation of an inventory profile that identified the best storage location for each part to improve the efficiency of order picking; improved workflow processes, leading to more efficient use of labor; and collaboratively enhanced the existing WMS system that enabled the company to streamline the operation further.

Ryder then keeps track of five key performance indicators (KPIs) on a monthly basis (the first four of which are also measures of waste): shipment accuracy, inventory accuracy, order fill rates, order cycle times, and budget performance [Ryder Logistics, 2011].

The point of these examples is to show that not only is the warehouse an ideal place to start a Lean retail and wholesale journey, but it can give you real results and a competitive advantage in the marketplace.

Lean Enabling Technology in a Warehouse

Technology in the distribution environment can be a true enabler of a Lean process for a retailer or wholesaler.

One of the main forms of technology found in this area is a WMS that controls the movement and storage of materials within a warehouse and processes related transactions such as product receiving, putaway, order picking, and shipping.

When a WMS is integrated with other technology already discussed, such as Universal Product Code (UPC) bar code scanners, RFID, and other software such as enterprise resource planning (ERP) systems, it helps to direct materials and personnel in the most accurate and efficient means possible, thereby eliminating waste. Ultimately, it ensures that the correct materials are getting to the right place, at the right time, in the proper quantities to satisfy customer orders, or demand.

One important benefit of a WMS is that it captures important data required for improving quality, minimizing variation, and improving workflows, allowing for the identification of waste.

A WMS records warehouse activities on a real-time basis to ensure that current information is always available in the form of reports, proactive notifications, and general information sharing. A paperless environment ensures improved accuracy (no transcription, reading, or writing errors) and the elimination of lost or damaged documents, thereby eliminating wasted effort and timing errors. Data entered are cross-checked automatically by the system to ensure validity. This accurate, real-time warehouse information can be used to facilitate improved management decision making and better collaboration with suppliers and customers.

As we now have a good understanding of the basic concept and tools of Lean as well as where to potentially apply them in retail and wholesale, we need to answer the question, "Where do we start?"

CHAPTER 11

Being Lean: Getting Started—Lean Assessments and Value Stream Mapping

So far, we have discussed many Lean concepts and possible applications in retail and wholesale. This leads to the question, "Where do we start the Lean journey?"

Lean Opportunity Assessment

In many cases, the best place to start your journey is to perform a Lean opportunity assessment (LOA) to help you identify the potential for improvement in your organization (see Fig. 11.1; a useful template, Appendix B, accompanies this book).

In an assessment of this type, you should identify and analyze various aspects of your company from a Lean perspective and rate them against best practices.

In our template (and there are many variations on this type of assessment), we look at various key performance areas and rate them anywhere from "traditional retail and wholesale" (i.e., "beginner") up to "world class retail and wholesale."

The areas recommended for evaluation within the office, warehouse, and store are:

- Internal communication
- Visual systems and workplace organization
- Employee flexibility
- Continuous improvement
- Mistake-proofing
- Quick changeover
- Quality
- Supply chain
- Balanced flow
- Total productive maintenance
- Pull system
- Standardized work

Lean Assessment Scorecard

	Internal communication	Score (1–5)
1	Management communicates with all levels of the organization on topics regarding organization goals and objectives at least twice per year.	
2	Employees are able to accurately describe the organization's goals and how their job contributes to the achievement of those goals.	
3	Employees receive feedback through a formal process concerning problems found in downstream processes or from the customer.	
4	Management encourages administrative, operations, and supply chain & logistics employees to work in groups to address performance, quality, or safety issues.	
5	Employees at the operations level understand and use common performance metrics to monitor and improve the production processes.	
6	Problems in the administrative, operations, and supply chain & logistics process are detected and investigated within 10 minutes of the first occurrence.	
7	The concept of value stream mapping is understood and all product famalies have been mapped and are physically segregated into the like process streams.	

Internal Communication Category Score = 0%

	Visual systems and workplace organization	Score (1–5)
1	The distribution center and office areas are generally clear of unnecessary materials, items or scrap, isles are clear of obstructions.	
2	The distribution center floor has lines that distinguish work areas, paths, and material handling isles.	
3	All employees are aware of good housekeeping practices and operators consider daily clean up and put away activities as part of their job.	
4	There is a place for everything and is everything in its place. Every needed item, tool, material container, or part rack is labeled and easy to find.	
5	Display boards containing job training, safety, operation measurables, production data, quality problems, and countermeasure information are readily visible at each production line or process and are updated continuously.	
6	Check sheets describing and tracking the top quality defects are posted and are up-to-date at each work station.	

Visual Systems Category Score = 0%

	Employee flexibility	Score (1–5)
1	Employees are given formal training before doing a job on their own. Few defects or productivity related slowdowns are attributable to new or inexperienced operators.	
2	Product/Component travel distances have been measured, analyzed, and reduced by moving equipment and work stations closed together.	
3	Equipment is "right sized" for the operartion/ process. They have to ability to change speed to match the TAKT time. No "monuments" are present in the process.	
4	Employees are cross-trained to perform other job functions and operators work in at least two different jobs each day.	
5	Processes and equipment are arranged to facilitate continuous flow of work through the distribution center.	
6	U-shaped cells have been designed and implemented to promote one piece flow where appropriate (eg: light assembly area).	

Operator Flexibility Category Score = 0%

Figure 11.1 Lean opportunity assessment sample section.

- Engineering
- Performance measurement
- Customer communication

Users then rate their performance in each category on a scale of 1 to 5 and compare it to Lean best practices to see where general opportunities for improvement exist.

Value Stream Mapping

While an LOA is a good first step, in order to actually identify, plan for, and make the improvements, you need a road map. The perfect road map for Lean is what is called a value stream mapping (VSM).

A value stream is the set of all actions (both value-added and non-value-added) that are needed to deliver a specific product or service from raw material through to the customer.

A VSM is typically the initial step your company should take in creating an overall Lean initiative plan. Developing a visual map of the value stream allows everyone to fully understand and agree on how value is produced and where waste occurs. It should be noted that in many cases, after companywide general Lean introduction training, many companies start instead (or at the same time) with a 5S program, which, as mentioned before, is a good foundation concept and is useful in transforming the company's culture from the bottom up.

Historically, businesses used flowcharts, which show the movement of materials; time function maps, which show flows and time frames; and process charts, which use symbols to show important activities. While these were adequate, the individuals who prepared them (often consultants) typically interviewed various parties engaged in a process in order to document it.

In many cases, this type of process mapping is used to enable business process reengineering (BPR), which is a total rethinking and redesign of a process. It was very much in vogue in the 1980s and 1990s, when companies were dramatically consolidating and outsourcing, resulting in large numbers of plant closings and layoffs in manufacturing.

This is in contrast to the concept of incremental change or continuous improvement, as it does not give any indication as to whether individual activities are value-added or non-value-added, especially from the viewpoint of the customer. That is one of the many improvements when using VSM, along with the fact that some data are also gathered to help to quantify the waste in an entire value stream, from customer to supplier.

Value Stream Mapping Defined

VSM is a team-based approach to mapping a value stream or process from beginning to end. It visually (and numerically) breaks the process down into value-added and non-value-added steps from the viewpoint of the customer.

Typically, you should start at the customer end and work your way back to the supply end. It is best to have a diverse team of employees (no more than 8–10) who are typically supervisory and management level (although frontline employees will be involved in the mapping process as well). It should typically take no longer than two days, one day for training and preparing the current-state map and the second day (usually a week or so later) to create the future-state map. The entire process, including implementation, should be completed in less than six months at most (if not sooner).

The team first maps the current state, while at the same time gathering ideas and input for an improved future state. The maps are drawn using standard symbols (easily found on the Internet) showing the flow of materials (shown at the bottom of the map) and information (at the top of the map). It should typically take no more than one day to map a process's current state, and this is really a "10,000-foot" view, so it should not be too detailed (but information should be validated).

The future-state map, based on the data contained in the current-state map, is the same value stream with any waste, defects, and failures eliminated. It is usually best to put some time in between doing the current- and future-state maps to give time for ideas to germinate.

Before starting on the current-state map, it is important to determine:

1. The value stream you are mapping
2. The product families that will be included
3. The takt time for the selected product family
4. The value stream manager
5. Your goals and objectives

Benefits of Value Stream Mapping

There are many benefits to VSM, especially when compared to the other methods described in the previous section.

A VSM shows the various connections between activities, information, and material flow that can affect the lead time of your value stream, allowing you to separate value-added activities from non-value-added activities and then measure their lead time.

It can also help employees understand your company's entire value stream rather than just a single part of it, as people tend to get tunnel vision or a "silo mentality," which is counterproductive when you are trying to implement Lean. VSM also provides a way for employees to easily identify and eliminate areas of waste.

At the same time, VSM can improve employees' decision-making processes by helping team members understand your company's current practices and future plans.

Other benefits include the following:

- Establishes priorities for improvement efforts.
- Focuses on no-cost or expenseable improvements.

- Provides a common language to talk about the processes.
- Is based on objective information.
- Forms the basis of an implementation plan.

Value Stream and Product Family

Typically, the first value stream selected for mapping may focus primarily on process improvement within the four walls of your company or facility, but will also be linked to key customers and suppliers. However, in the case of retail and wholesale, the focus may be broader from the start, as the supply chain connects us with our suppliers and our customers. We may be mapping a process in our store, warehouse, or distribution center (DC), or a process such as purchasing and/or transportation, which is directly linked to our suppliers and vendors.

In any case, it is important that we first narrow the scope of our VSM to a specific product family. A product family consists of products or services that have mostly the same steps. That does not mean that all the products or services have to have all of the same steps (just mostly the same steps!).

Takt Time

As discussed earlier, takt time, which is the demand rate for a product or service family, shows the pace for each activity in the value stream and is useful in identifying bottlenecks that constrain or limit capacity. In a VSM, the takt time is usually calculated in minutes or seconds (i.e., the need to ship so many units every so many minutes or seconds).

Value Stream Manager

Every team needs a quarterback, and each value stream needs a value stream manager. For product ownership beyond functions, you should assign responsibility for future-state mapping and implementing Lean value streams to line managers who have the capability to make changes happen across functional and departmental boundaries. Value stream managers should make their progress reports to the top manager on the site.

Goals and Objectives

It is a good idea to create a "steering committee" to manage and support the Lean journey. This is especially true in the case of VSM, as it sets the direction for the company and drives individual kaizen events. It is also critical that the goals of the VSM and subsequent kaizen events be tied to the company's strategic and operational goals. This usually is not too hard to do as long as the company has established these goals. Typically, they can include metrics covering quality, manufacturing, supply chain and logistics, store operations, and customer service, to name a few areas. Specifically, they may measure things like inventory turns, on-time and accurate shipments, cases per hour handled, and the like.

In a more general way, when setting goals and objectives, the team should be asking questions like:

- What does the customer want in terms of the cost, service, and quality of our products or services, and which objectives and goals established by our company meet these market needs?
- What processes greatly affect the performance of our products and services?
- Who needs to support this effort?

Steps in Creating a Current-State Value Stream Map

While there are standard "icons" that are used in VSM (easily found on the Internet), there really is no "standard" step-by-step methodology for the actual mapping process. In general, you should:

1. Have individual team members do a "quick and dirty" map of the major processes in a value stream by themselves, and then, as a group, compare maps and reach a consensus. This should typically just include the individual processes (from customer to supplier) to make sure that none were missed. It is always useful to draw the major activities on a whiteboard once they have been agreed to at this point.
2. Assemble data collection forms (to be discussed shortly), pencils, "yellow stickies," and a stopwatch for collecting data.
3. Select a product family (or service) to map. Conduct a quick tour of the value stream to view the end-to-end material and information flows, making sure that you have identified all the component flows. (It is usually best to start in shipping and work your way back.) Always let operators know in advance about the tour so that they don't work any faster than or differently from normal. During this time, it is important to observe and ask questions and to start thinking about the "future state." These ideas should be captured on the yellow stickies.
4. Identify representative customers for the product and gather data about typical order quantities, delivery frequencies, and number of product variations (these will be used to develop takt time).
5. Assign team members individual value stream activities for which to collect data, using the data collection form.
6. Begin mapping the detailed value stream (using Lean icons), starting with customer requirements and going through the major production activities. The result is a current-state value stream map (information flow at the top, material flow in the middle, and a lead-time chart at the bottom of the VSM, showing value-added and non-value-added production lead times).

7. Review the map with all the employees who work in the value stream you have mapped to ensure accuracy (see Fig. 11.2 for a current-state VSM example for a national parts distributor).
8. Review the map with upper management before proceeding with the future-state mapping process.

Data Collection

When doing VSM in retail and wholesale, in some cases, it may be more like a "traditional" VSM in manufacturing, where you view the inventory that is built up between processes as non-value-added and convert the quantities to days or even minutes using the takt time.

However, in the office, the warehouse, and sometimes the store, it may be more important to look at inventory (and information) "dwell" time between processes as a better indication of waste.

In a warehouse or DC it is important to see how long inventory sits between receiving and putting away, picking and staging, staging and shipping, and so on. It is always a good thing to also measure the amount of inventory between steps, but in many cases, the DC may have no control over the amount of inventory (that's a VSM and kaizen event for another day with corporate buyers and purchasing managers!). It really depends on how far upstream you take your value stream.

In the office, dwell time between steps is important, as the longer information such as orders sits, the longer the order-to-cash cycle becomes.

The type of information gathered to develop a VSM also differs somewhat in an office, DC, or store (see Fig. 11.3). The reality of mapping in a retail and wholesale environment is that it is usually a combination of an office, store, and warehouse (including transportation) value stream.

In the store, the issue may be days' supply of inventory in the back-room storage area or on the store shelves, in which case a more traditional VSM model (like that in manufacturing) can be used. In many cases, though, we may also be looking at more of a service process for VSM, such as product returns, customer service, checkout, and value-added services such as paint mixing and the pharmacy, in which case dwell or waiting time may be most important.

VSM can still be used in these areas, but it may sometimes be too broad for a narrower service function such as those just mentioned. In those cases, one should consider using service blueprinting (mentioned in Chap. 9), as customer interaction is most important in those areas, and service blueprinting helps us to focus on the customer and is an important variable in the design process.

Process charts (discussed in Chap. 6) are also very useful in mapping a store operations process, as they are designed to provide a more detailed view of a process, including measurements such as time (value-added and non-value-added), delay, distance, and storage to make the chart more meaningful from a Lean perspective.

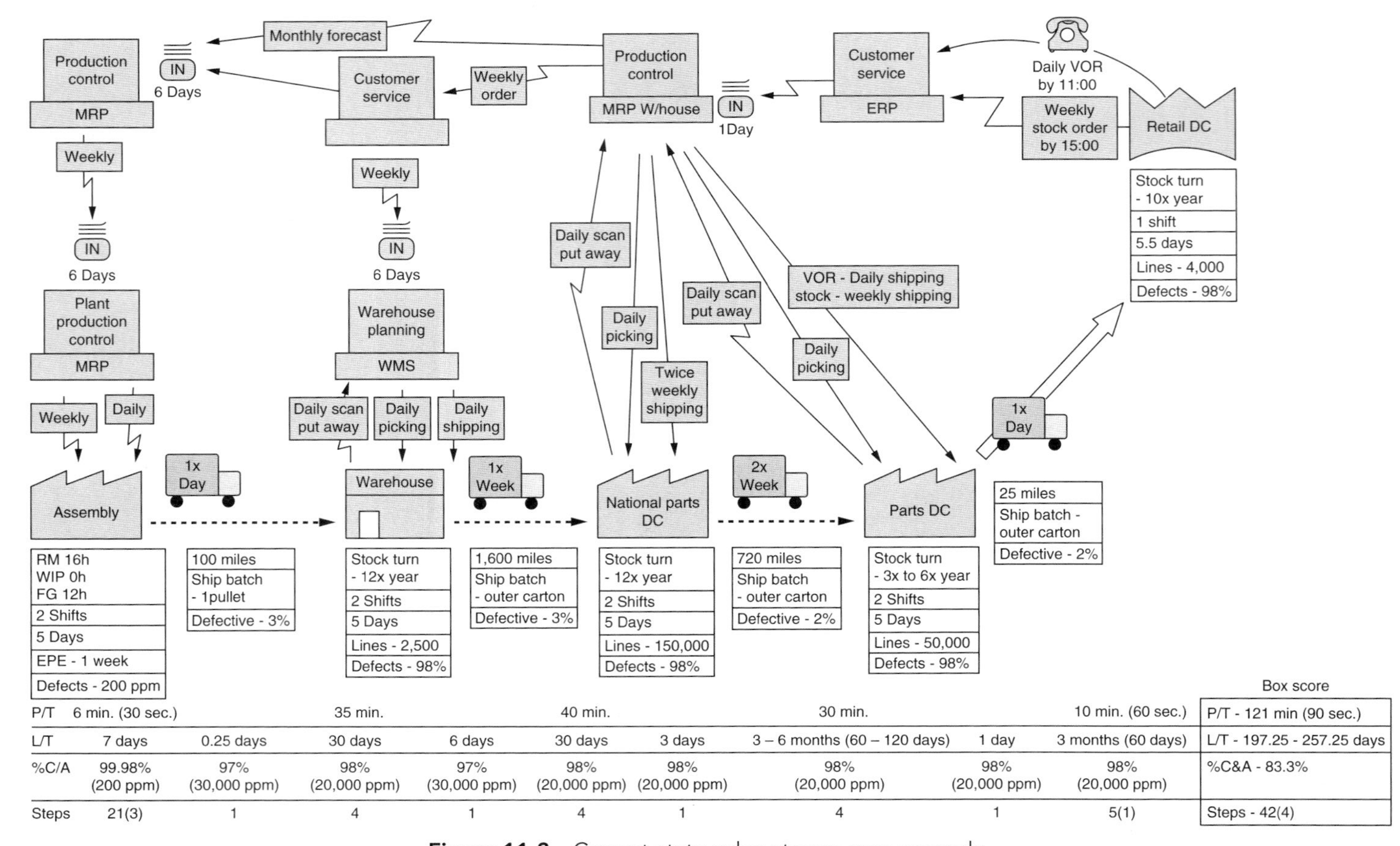

Figure 11.2 Current-state value stream map example.

Retail and Wholesale Value Stream Mapping
Data Collection Sheet

- Process time
- Available time
- Setup time
- Lead time/turnaround time
- Typical batch size
- % Complete and accurate information (% C&A)
- Rework/revision (e.g. design changes)
- Number of people involved
- Reliability (e.g. system downtime)
- "Inventory" – queues of information (e.g. electronic, paper) and/or physical inventory (Raw, WIP or Finished Goods)

Figure 11.3 Retail and wholesale VSM data collection sheet.

The data collection form shown in Fig. 11.3 gives you a good idea of the type of information required to create a VSM. Once the data are collected and displayed on a current-state VSM, along with information flows, the team can get a good idea of where improvement opportunities exist.

The actual layout of the current-state VSM can be seen in Fig. 11.2.

The value stream is mapped from customer to supplier, as previously mentioned.

Data boxes are completed with the information gathered for each process or activity in the middle of the map (with either inventory or dwell time between the activities), information flow shown at the top, and individual and systemwide lead times (value-added and non-value-added) at the bottom.

As the team members discuss and reach consensus that the current-state map is accurate, they start to identify areas for improvement with the goal of reducing non-value-added activities. These opportunities are sometimes highlighted by "kaizen bursts" around affected activities in the current-state map.

In our example of the current-state map for a national parts wholesale distributor process, the team found many opportunities to reduce non-value-added steps, focusing on reducing lead times and inventories (discussed in more detail a little later in this chapter).

Future-State Value Stream Map

The future-state map attempts to create a flexible, agile system that quickly adapts to ever-changing customer needs, eliminates waste, minimizes handoffs and "silos," and triggers resources only when needed (see the example in Fig. 11.4).

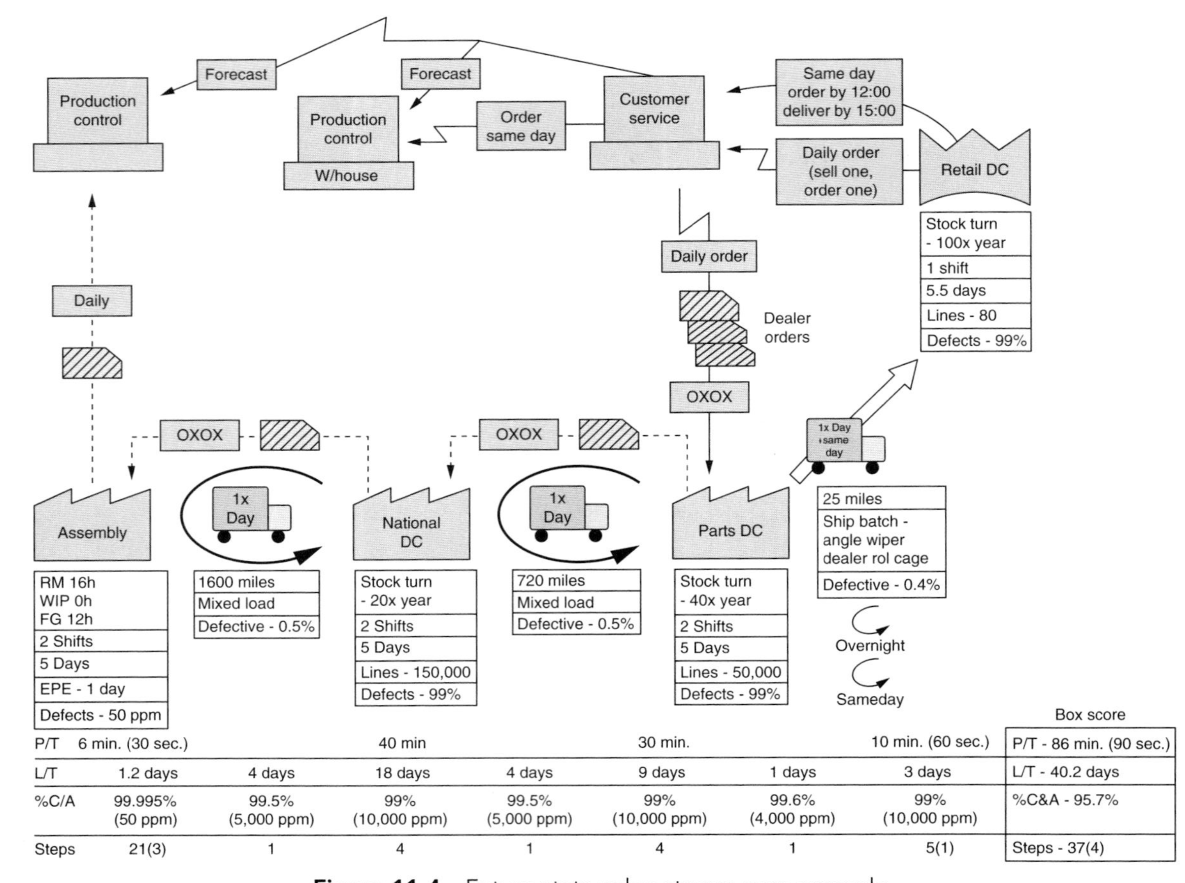

Figure 11.4 Future-state value stream map example.

As mentioned previously, the team members should have developed some thoughts on improvement during the information-gathering phase of the current-state map. When they start to actually "put pen to paper" (or on a whiteboard) for the future-state map, they should ask some questions, such as:

- What does the customer really want or need?
  This should include determining what "service level" the customer needs, response or turnaround times, the required quality level of the output, the expected demand rate and variation, and the resources required to meet the demand rate.
- How often will we check our performance relative to customer needs?
  This helps to define the frequency with which the system will be reviewed to verify that it is satisfying customer needs and meeting the targeted cycle time or service level, and how we will check its progress.
- Which steps create value and which are waste?
  We should challenge every step and determine what is really needed by the customer and what can be done differently (or not at all). We also need to know if existing controls and administrative guidelines are appropriate and what knowledge and skills are truly required to perform the step(s).
- How can we flow work with fewer interruptions?
  We need to understand whether we have continuous flow in areas such as customer service (e.g., from order to invoice), order processing (e.g., design-to-order), and warehouse and distribution (e.g., identifying bottlenecks in the process).
- How do we control work between activities?
  It is always useful to ask employees how they know what to do next when they have finished with a batch of work.
- Can we better balance workload and/or different activities, and how will work be prioritized?
  We need to know whether the "mix" (i.e., the order, product, or service type) affects the ability of the system to flow or affects the responsiveness of particular steps in any way, and whether the "volume" (i.e., demand variation) affects the system.
- Finally, what process improvements will be necessary, and which are most important?

The answers to these questions will drive the future-state map and implementation plan. This is also where the team can consider applying many of the basic and advanced Lean concepts discussed earlier in the book, such as layout, visual workplace, setup reduction, just-in-time (JIT), pull/kanban, work cells, and so on.

Where to Look

Overall, the focus should be on eliminating waste while trying to improve continuous flow. There should be an emphasis on shifting from a push to a demand-pull process based upon customer demand (using concepts such as takt time, load leveling, process balancing, and one-piece flow) to minimize inventory. You should target more visibility in both the demand and the supply chain to help manage service and costs.

As purchasing can account for 50 percent or more of total costs, Lean tools can be implemented, such as visual material management (e.g., kanbans and simple bin replenishment), vendor-managed inventory (VMI) systems, and supplier reviews or assessments to see how Lean your suppliers really are. There should be an emphasis on partnering, collaborating, and long-term alliances, with sharing of cost and technical data and, in some cases, risk. All of this will require mutual trust.

In warehousing, waste can be found throughout, including:

- Defective products, which create returns
- Overproduction or overshipment of products
- Excess (or inaccurate) inventories, which require additional space and reduce warehousing efficiency
- Excess motion and handling
- Inefficiencies and unnecessary processing steps
- Excess transportation steps and distances
- Waiting for parts, materials, and information
- Inefficient (and manual) information processes

In a retail store, it is helpful to think of waste in terms of the five points of transition for a customer (i.e., enter, seek, find, select, and transact), as discussed in Chap. 1. The types of waste to look for in each are:

The final future-state map should be reviewed with the team, then presented to upper management for its approval and support.

Five Points/ Wastes	Motion	Waiting	Excess Information	Excess Inventory	Defects/ Errors	Over-processing	Trans-portation
Enter	X	X	X				
Seek	X		X	X	X		
Find	X	X			X	X	
Select	X		X		X		X
Transact	X	X			X	X	

In our example of the future-state map for a wholesale supply chain function (previously shown in Fig. 11.4), the national parts wholesale distributor team determined that it would be an improvement for the distributor to go from a push type of system to a demand-pull system using kanban cards at both its local and national DCs, placing replenishment orders more frequently in smaller order sizes. While this may increase transportation and material handling costs somewhat, it reduces inventory carrying costs and helps the distributor to become more flexible.

The distributor team also decided that it would be a good idea to start a VMI program with the manufacturer of this product, which not only would allow the firm to use a kanban system for the product's replenishment, but would improve inventory turns, reduce out-of-stocks, and pass more responsibility (and information) on to the supplier, creating a long-term stable relationship.

Both of these improvements would involve improved communication and collaboration with retailers and manufacturers.

At some point, the future-state map becomes the current-state map. As a result, a year later, the wholesale distributor team began formulating a plan to bypass (or eliminate) its regional DC and change a section of its local DC (or possibly a new facility) to a distributor "cross-dock operation" for some large retail customers (see Fig. 11.5).

Cross-docking refers to moving product from a manufacturing plant and delivering it directly to the customer, with little or no material handling in between. Cross-docking not only reduces material handling, but also reduces the need to store the products in the warehouse. More specifically, distributor cross-docking occurs when the distributor consolidates inbound products from *different* vendors into a mixed product pallet, which is delivered to the customer store when the final item is received. As it is a mixed pallet with minimal quantities of each item, very little, if any, stock needs to be kept in the back of the store.

The distributor team in our example also realized how important it was to collaborate and share information (point of sale (POS), store inventory balances, and so on) with the customer as well as the need to set up Collaborative Planning, Forecasting and Replenishment (CPFR) teams with the key retail customers to

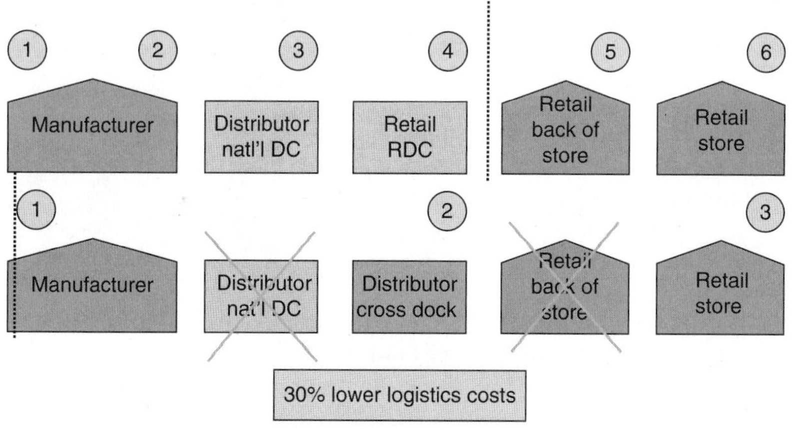

Figure 11.5 Conceptual longer-term future-state map.

improve forecasting, minimize out-of-stocks, and improve customer service levels. The CPFR program has the added benefit to the distributor of creating more of a long-term partnership with the customer.

Implementation Plan

The final step (and really just the beginning) is to prioritize and plan for the improvements identified in your future-state map to be implemented in the next three to six months. There is usually some "low-hanging fruit" that can be done quickly and easily. The harder improvements, which may take some capital improvements and more time, also need to be prioritized and planned for (see Fig. 11.6).

It is a good idea to tie the future-state map to the overall business objectives, as mentioned previously. It is also sometimes best to break a future-state map into easier-to digest and easier-to-deliver loops. For example, you may create separate customer, warehouse, and supplier loops.

To ensure success, you will need leadership and support from upper management, a company business strategy that includes Lean, development of a culture that emphasizes a continuous improvement mentality, and ways to ensure that Lean becomes part of everyone's job description and is discussed on a regular basis.

Additionally, it is always a good idea to benchmark key metrics within your industry and outside your industry and to always question the status quo.

Lean Teams

In many ways, the Lean journey by itself is a form of team building, especially when you use a simulation training game early in the process to help trainees transfer the general concepts and tools into hands-on learning.

In any case, much more progress can be made and maintained if Lean is part of everyone's job, rather than that of just a few. While it is important to have a Lean structure, with Lean champions, Lean coordinators, and even, in some cases, Lean subject-matter experts (in areas such as VSM, 5S, and so on), it is more important and effective to have everyone involved from top to bottom.

Companies in which Lean becomes part of the culture tend to get greater, longer-lasting results. While those that just look at it as the fad of the month may get some good results, they usually go back to doing things the way they were done before the training. What you really want to do is develop a culture that consistently defines and solves problems utilizing Lean tools.

Yearly Value Stream Plan

Date:	
Facility manager:	
VS manager:	

Signatures			
Plant mgr	Union	Eng	Maint

Product family business objective	VS future state loop	VS objective	Goal (measurable)	Monthly schdule												Person in charge	Related individuals & depts	Review schdule	
				1	2	3	4	5	6	7	8	9	10	11	12			Reviewer	Date
	1																		
	2																		
	3																		

Product family:

Figure 11.6 Value stream plan.

There are many reasons to work in teams, and in fact, doing so is pretty much the norm these days, including everything from new product teams to quality teams and beyond. Teams are a great way to share ideas, and they create a support system for their members. They also use the skill sets of all of the members, as the sum of ideas is usually better than an individual person's ideas. Besides, it is more fun to work in teams, and as a result, implementation is usually easier.

When creating teams and a culture of teamwork, it should be in a risk-free environment, but it should still be disciplined in terms of the process and the rules. Trust is a critical factor, as is selecting the right people for the team.

I always prefer a mix of people from a variety of areas (e.g., production, supply chain, logistics, engineering, sales, and so on) to gain a common understanding and to avoid silo thinking. I also prefer to take volunteers first.

Team Charter

It is always good to formalize things with a team charter (Fig. 11.7) so that everyone knows his or her roles and responsibilities.

The Team Makeup

You should make sure that you have a fairly high-level sponsor or advisor for the team to help break through any roadblocks along the journey, as well as being able to arrange for support for the team during the event(s).

Team Charter

Date: ____________

Targe area: ____________

Target area supervisor/manager:

Mission/purpose: ____________

____

Team name: ____________

____

Champion/facilitator:

Leader: ____________

____

Team members and roles:

Recorder: ____________

Time keeper: ____________

Members: ____________

____

Scope of project: ____________

Critical success factors
(including metrics): ____________

____

Figure 11.7 Team charter.

The team leader is kind of like the quarterback, determining session objectives, the process to be followed, and the agenda. The team leader should meet with the facilitator to review session objectives and processes. The traits to look for in a leader include any previous success as a leader, some knowledge of Lean Enterprise (preferably hands-on experience), and someone who is comfortable working in the targeted area(s).

The consultant/trainer should be more of a facilitator. I typically deliver Lean concepts and application training, but I usually end up doing consulting as well to help steer the team in the right direction in general, as well as to help come up with and organize specific improvement ideas.

The actual team members should be a blend of people from inside and outside the area, as mentioned previously, and they should understand the target area (they may work in it, or they can learn about it) and be open to thinking about doing things differently.

When it comes to meetings and their rules, it is important to set them early and try to adhere to them as much as possible. They should include basic things such as showing up and starting on time, being prepared, listening attentively, participating, and so on, as well as individuals assuming responsibilities and supporting any group decisions.

It can be very useful to have a place on the whiteboard (or a large piece of paper) to use as a parking lot for good ideas that may be off-topic at the moment, but that have longer-term potential later.

Kaizen Events

Kaizen is loosely translated (or commonly known) to mean continuous improvement. So a kaizen event is a team-based continuous improvement project. The goal of a kaizen event is process improvement through the elimination of waste at all levels of the process. Typically, prior to the actual event, there is a period of training in general Lean concepts and applications, as well as additional training in specific tools, such as setup reduction, work cells, and so on.

As mentioned earlier in the book, the first kaizen event is usually either 5S workplace organization or implementing the results of a VSM project. However, you can also create a kaizen event based upon feedback from your team members after some basic training.

For example, after some basic Lean training, the general manager at a client facility that functioned as both a private DC for a cosmetics contract manufacturer and a third-party logistics services (3PL) facility for some major cosmetics retailers such as Victoria's Secret and Bath & Body Works had his team do some brainstorming to come up with ideas for kaizen events. We handed out a form similar to the one shown in Fig. 11.8 and asked the individuals to come up with improvement ideas.

The team came up with many good ideas, including one idea for improving the cycle-counting process. The facility's current cycle-counting process was done

Cost Reduction Kaizen Implement

Department: ____ Process for kaizen: ____________ Kaizen #: ______
Cost center: ______ Date: __________
Approvals: Lean champion: ___ Maint: ___ Controller: ___ GM: __________

1) Current situation	3) Solution activity	
2) Analysis	4) Cost reduction	(Total savings:)
	Current	Proposed

Figure 11.8 Cost reduction kaizen implementation form.

manually, even though the DC had radio-frequency (RF) scanning capabilities for receiving, picking, and so on. Time studies showed that the facility was wasting an average of 108 minutes each day doing data entry and 54 minutes in wasted travel time to the office. The solution was to use RF devices for cycle counting, which would not only significantly reduce these wastes but also cut down on data entry errors.

In general, a kaizen event is appropriate in a number of situations, including when there is a need for an urgent solution and when there are competitive issues, customer service issues, cost issues, or bottlenecks.

A major difference between kaizen events in America and in Japan is that in the United States there seems to be more of a rush to get results, in contrast to the slow but sure way of continuously improving a process in Japan. That may be one reason that many Lean initiatives in the United States fail. The companies may make short-term gains, but they don't stick with it for the long haul the way Japanese companies do.

Kaizen Event Management

When planning for a kaizen event, there are some general steps that are good to follow:

1. Prepare for the event by selecting the targeted area and the team.
2. Define the scope and goals of the event with the team members.
3. Train the team in various Lean concepts and relevant applications.
4. Walk the kaizen area with the team members.
5. Collect data on the kaizen event area (this varies depending on whether it is for a VSM, 5S, or some other specific tool to improve the process by removing waste).

6. Brainstorm ideas as a group.
7. Prioritize the top ideas in terms of value to the customer.
8. Form subteams to implement the ideas.
9. Keep track of progress and check results.
10. Develop, review, and update employee job instructions where needed (adding standardized work wherever possible).
11. Develop an action plan for the remaining ideas.
12. Regularly report the plan, its progress, and its results to management.
13. Recognize the team and communicate results to the entire organization.
14. Follow up on open action items.
15. Measure area improvement relative to goals and objectives.
16. Disband the team when the kaizen is finished.

From an upper-management perspective, whether from a steering committee or a Lean champion standpoint, in addition to each kaizen event being organized similar to this list, it is key that there is some kind of kaizen board to keep track of the kaizen schedules and progress.

A kaizen board is an ideal tool for making the events visible. It can be a simple cork- or whiteboard. This is useful for controlling progress, sharing information, and keeping motivation going.

After some general overview Lean training, a VSM is a great first step and a valuable tool for continuous improvement. It serves as a "road map" to the future and is a true foundation for the entire Lean journey.

CHAPTER 12

Staying Lean: Leadership, Culture, Teams, and Key Success Factors

Lean is both a philosophy and a process improvement system. While its use in retail and wholesale so far has been on a more limited (supply chain) basis, it is important to take a more "holistic" view of it and permeate your entire organization with both the will and the way to make it work.

To do this, it takes a major effort by everyone involved, from the president to the store or warehouse workers, for it to be successful. In this chapter, we will describe ways to ensure Lean's continual implementation, monitoring, and long-term success.

Building a Lean Organization

Being successful in building a Lean organization requires the commitment and involvement of managers, employees, partners, and suppliers in order to continually monitor and improve a process. That being said, as an organization, we need to consider the following before undertaking any large change-oriented endeavor to be successful [Rummler and Brache, 1995]:

1. *Connect with strategic issues*—During process analysis, we need to pay attention to core processes, competitive priorities, the impact of customer service contact and volume, and strategic fit.
2. *Involve the right people in the right way*—Make sure the process analysis involves the people who are performing the process or those who are closely connected to it as internal customers and suppliers.
3. *Give design teams and process analysts a clear charter and hold them accountable*—Management should set expectations for change and maintain pressure for results. Don't fall prey to "analysis paralysis," where nothing ever seems to get done.
4. *Evolution or revolution?*—Lean is a long-term transformational process. Unless we are expecting radical reengineering (which may occur in some

cases), we should set goals that consider the cumulative effort of many possibly smaller improvements.

5. *Consider the impact on people*—The changes should align with the attitudes and skills of the people who are implementing the redesigned process. We need to understand and deal with the people side of process changes.
6. *Give adequate attention to implementation*—We need to use sound project management practices with a strong team leader.
7. *Have an infrastructure for Lean continuous improvement*—This includes not only the Lean champions, coordinators, subject-matter experts (SMEs), and necessary supplies, but also making sure that there are measurements in place to monitor key performance metrics over time.

The importance of involving employees in a Lean journey and considering its impact on them can't be overemphasized. Evidence of that can be found in the results of:

> A [2007] study of 90,000 workers in 18 countries conducted by Towers Perrin on the impact of employee engagement [that] states the following: "Firms with the highest percentage of engaged employees collectively increased operating income 19% and earnings per share 28% year-to-year. Companies with employees who were less engaged experienced operating income declines of 33% year-to-year and 11% in earnings." Their study also states, "It's impossible to overstate the importance of an engaged workforce on a company's bottom line. The study demonstrates that, at a time when companies are looking for every source of competitive advantage, the workforce itself represents the largest reservoir of untapped potential."
>
> Lean beliefs and Lean work processes enable employees to "rise out of the generic and into the particular and skillfully engaged." By doing this they have given us the ability to greatly increase the productivity and profitability of 21st century workplaces, of all types, definitely including retail work environments.
>
> Knowing this, how can your work environment be transformed into one that drives high levels of both employee and customer engagement? [Friesen, 2011]

The answer is to develop a Lean culture for continuous improvement that is integrated with Lean concepts, tools, and planning, which is not an easy task.

Organizational Culture

First, we need to define what an organizational culture is in the first place. In general, it's the values and behaviors that contribute to the unique social and psychological environment of an organization. It includes the organization's expectations, experiences, philosophy, and values that hold it together, and it is expressed in the organization's self-image, inner workings, interactions with the outside world, and future expectations. It is based on shared attitudes, beliefs, customs, and written and unwritten rules that have been developed over time and are considered valid. This is also often called corporate culture, and it's exhibited in:

1. How the organization conducts its business and how it treats its employees, its customers, and the wider community.
2. The amount of freedom employees have in decision making, developing new ideas, and personal expression.
3. How power and information flow through the organization.
4. The extent to which employees are committed to organizational objectives.

Culture affects the organization's productivity and performance, and provides guidelines on things like customer service, quality, and employee attitudes. Organizational culture is unique for every organization and is one of the hardest things to change.

Lean Culture

One way to define Lean culture is that it is an organizational environment in which values and behaviors are aligned with the guiding principles of Lean thinking.

With that in mind, I think it's a good idea to then use the four guiding principles of the Shingo model (based on the Lean management approach and model taught by Dr. Shigeo Shingo that used three levels of business improvement: principles, systems, and tools/techniques) to support a Lean culture:

1. Lead with respect and humility.
2. Establish enterprise alignment.
3. Pursue continuous process improvement (CPI).
4. Deliver customer-focused results.

Additionally, it's a good idea to:

- Have a never-ending commitment to excellence and under no circumstances accept mediocrity.
- Invest in coaching, developing, and mentoring employees.
- Foster good communications (up, down, and sideways in the organization).
- Develop leaders into "owners" rather than "renters."
- Encourage an environment of creativity and critical thinking.
- Encourage teamwork and collaboration.

A good example of a Lean culture that starts at the top is that of Tesco, a British multinational grocery and general merchandise retailer that is the second-largest retailer in the world measured by profits (after Walmart).

> [Under former president] Sir Terry Leahy's leadership Tesco's turnover quadrupled from £16 billion to an anticipated £64 billion, profits increased by six times from around £532 million when he started to an

estimated £3.2 billion in 2010, and earnings per share increased by more than three and a half times.

Tesco's motto, said Leahy is, "better, simpler, cheaper." Everything they do, Leahy explained, should be "better for the customer, simpler for the staff, and cheaper for Tesco." This culture of simplicity also encourages lean thinking [simplicity and Lean thinking are two of the ten management lessons that have guided him]. Businesses should always think lean because even in a successful business there is still "an awful lot of waste."

Encouraging others to have ideas is for Leahy a crucial part of any leader's role. "What is important is what a leader causes other people to do," he told the London Business Forum. Creating other leaders throughout the organisation ensures not only its longevity but also motivates people to reach big targets." [London Business Forum, 2010]

Key Success Factors

There are a number of *key success factors* (KSFs) that are critical to establishing a Lean culture and a successful Lean journey in any type of organization.

Management Leadership and Support

It sounds simple, but no type of change can be successful without top management actively driving and supporting the change with strong leadership (as seen in the Tesco example just given). When companies do employee training using grants, management often seems to look at it as "free" training, without considering the potential long-term commitment required to get the potential benefits of the training. That type of thinking is rarely successful for other than short-term results. Management has to be in it for the long haul; otherwise, it is doomed to fail. In fact, it is a good idea that your overall company business strategy includes Lean.

There has to be a commitment on the part of *everyone* in the organization, not just management, to make it work. Lean and continuous improvement in general should become part of everyone's job, from the job description to the review and reward system. A small amount of time should be set aside during every meeting to discuss Lean projects and progress. Everyone should be encouraged to question the "status quo."

Lean Training

First of all, it's important to train the entire organization to ensure that everyone understands the Lean philosophy and has an understanding of all the concepts and tools. It can be useful to have an outside trainer and facilitator come in to lead kaizen events, especially in the early stages. As they used to say about consultants, "They borrow your watch to tell you what time it is." I delivered some 5S training and kaizen events at the truck maintenance facility at a 1.5 million-square-foot distribution center for a major toy retailer. Because of rapid growth of the fleet and limited space, the facility's storage area had become quite a mess. While the company was aware of this, it took someone from the outside to help people look at things differently and understand the benefits of workplace organization. The end result was more storage space, which was desperately needed (and it was much more organized, so it was easier to find spare parts and supplies).

Lean Infrastructure

A carefully designed Lean infrastructure will keep improvement projects focused on those areas that matter most to the organization. The infrastructure should also provide techniques for tracking the progress of improvement efforts to enable the organization to determine what is working and what is not working. Corrections can be made when they are needed, and successes can be celebrated when they occur.

An organization should have methods to:

- Select and prioritize meaningful process improvement opportunities.
- Track and manage the overall portfolio of kaizen events.
- Communicate Lean kaizen results to the broader organization.
- Monitor and share key metrics that track the progress of Lean efforts throughout the organization.
- Define clear Lean roles and responsibilities for all employees.
- Establish a steering team to provide oversight to CPI efforts.

It's always a good idea to designate a *lean champion* to spearhead continuous improvement activities. In many cases, those companies that really believe in the benefits of Lean hire full-time Lean champions. When doing so, you should try to find a good, experienced change agent as the champion. The more structure you have to accomplish your Lean goals, the better your chances for success. Some companies also assign Lean coordinators and even SMEs. The coordinators are responsible for various kaizen events, and the SMEs may be trained to manage specific types of events (5S, value stream mapping (VSM), and so on).

It's usually the Lean champion's responsibility to create a kaizen agenda and communicate it to operators and others in the organization who are members of the empowered teams.

To get started, it's usually best to pick an area that is visible, yet manageable in size and that can be dramatically improved, such as a small supply storage or work area with a lot of traffic. This way, the organization can see both that you're serious about Lean and how it can make a difference. As mentioned before, after giving everyone some general Lean overview training, some companies go right to VSM and others start with 5S events. Which way a particular company goes depends on its size and the industry it's in, among other things. In any case, it's important to start mapping value streams to identify opportunities and to begin as soon as possible with an important and visible activity. While doing this mapping, it is always helpful to benchmark or compare data within your industry and outside your industry to help you prioritize improvements.

It's important that you become as Lean as possible within your company's "four walls," and then expand to customers and suppliers. In many cases, if you don't eventually partner with suppliers and major customers on the Lean journey, the full benefit of Lean is never achieved, and in some cases, you are just passing your inefficiencies on to your suppliers, making them less efficient.

One Lean client we had was partially funded by a major (large) customer who had been through a Lean journey itself. The customer had reached the point where it wanted its major suppliers to become more efficient as well. There were, however, "strings attached"—the supplier was responsible for improving specific metrics, such as on-time delivery, lead times, and the like. That way, it was a "win-win" situation for both the customer and the supplier.

Teamwork and Lean

Teamwork is fundamental to competing in today's global environment in general, and to Lean in particular, and should be part of any Lean retail strategy, as touched upon in Chap. 7. Teams help to support a process that defines and solves problems using Lean tools.

These days, most employees have been on some kind of team, whether it be a management team, a quality circle, a new product team, or something else. It's pretty much the "way of the (business) world" these days. In fact, it seems that teams are more the norm than the exception in today's environment.

Making Teamwork Happen

To make teamwork happen, executives must communicate the clear expectation that teamwork and collaboration are expected from the group. Management must talk about and identify the value of a teamwork culture, and must recognize and reward it. Over time, stories and folklore develop that people discuss within the company that help to emphasize teamwork.

Most successful teams have a large degree of employee empowerment, or the authority to make decisions on their own. In many cases, teams are "self-directed," which gives the members this added sense of empowerment. In order to ensure success, management must provide adequate support and training and must establish clear objectives and goals. There's always the "What's in it for me?" question, so the idea of financial and nonfinancial rewards is important. From a simple "pat on the back" or treating the team to lunch to actual financial rewards can be useful and meaningful.

In general, it can sometimes be difficult for managers to give up control to the team, but as long as they make sure that there are adequate controls in place, it can be successful.

When teams are successful, organizations and employees can benefit from an improved quality of work life, increased motivation, improved satisfaction, better quality and productivity, and lower turnover, among other things.

On the other hand, there are higher costs involved as a result of increased wages, training, and capital costs needed to be successful.

Elements Needed for Sustainability and Success

So, how do we integrate culture, leadership, teams, and training into one coherent train of thought? One way is to use a leadership model to guide the way, such as the one seen in Fig. 12.1.

Leadership Model

The steps involved in the leadership model would be:

1. Readiness check
    - Identify strategic and organizational priorities.
    - Assess the leadership structure and development needs.
    - Identify knowledge deficits and learning competencies.
    - Set the deployment pace based on the organization's needs.

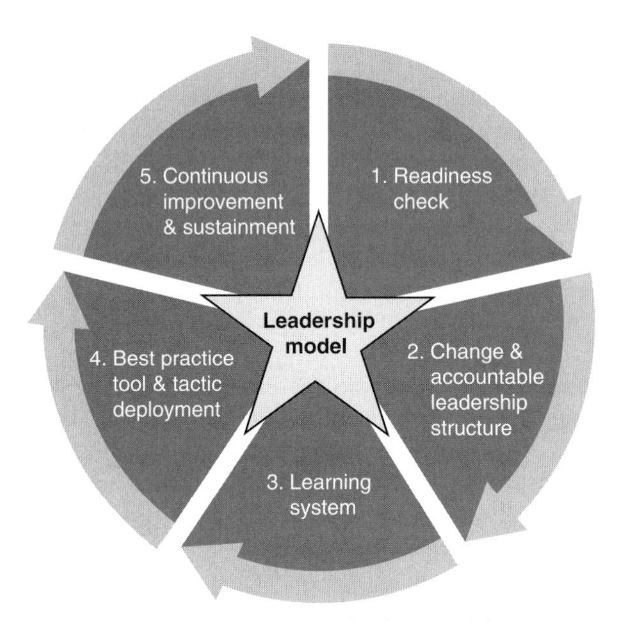

Figure 12.1 Leadership model.

- Allow for the time and resources that are required to permit change to happen.
- Don't assume that everyone is at the same level of readiness.

2. Change and accountable leadership structure
   - Set clear expectations for yourself and others.
   - Take personal responsibility for performance and results.
   - Follow through and "do what you say you'll do."
   - Use standardized continuous improvement action plans (what is the action, who is accountable, when is the targeted completion date, and how will you measure it?).
   - Be transparent in all goals and measures.
   - Enforce consistent performance tracking and responsiveness.
   - Consistently be a role model for values and behaviors.
   - Be *actively visible*: the leadership should always be readily available, especially when problems occur (i.e., "management by walking around").
   - Make change a standing agenda item in all meetings and venues—it needs to be seen as important.
   - Be willing to ask for support, even as leaders.
   - Provide real-time coaching in the moment.
3. Learning system
   - Deploy widespread learning about the change process prior to initiating any improvements.

- Translate process improvement into a language that workers can understand.
- Provide structured and practical learning with applicable examples.
- Incorporate just-in-time learning into improvement efforts.
- Ensure that training and education are widespread and adapted to the senior leader level.

4. Deploy industry best practice tools and tactics
   - Focus on using the "right tool, for the right purpose, at the right time."
   - Attack the broken system, not the people.
   - Focus on effective problem solving and root cause analysis.
   - Integrate leadership and behavioral accountabilities with process improvement tactics.
5. Continuous improvement and sustainment
   - Understand the changes that need to be made.
   - Take an active part in the change process.
   - Don't underestimate the time needed for change to become widespread.
   - Plan for shortfalls and missteps—and learn from them.
   - Enable associates to be change agents for what needs to get done.
   - Establish a foundation and a shared need for change before improving processes.

Perhaps the hardest part of any type of process improvement project is sustaining the changes that were made.

An IBM study on organizational change [Jørgensen et al., 2008] found that only 41 percent of projects were considered successful in meeting project objectives within the expected time. The authors concluded that, not surprisingly, success depended more on people than on anything else.

The study identified four common factors that helped practitioners address their greatest project challenges. They were:

1. *Real insights, real actions*—Have a realistic awareness and understanding of the upcoming challenges and complexities, and follow with actions to address them.
2. *Solid methods, solid benefits*—Use a systematic methodology for change that is focused on outcomes and aligned with formal project management methodology.
3. *Better skills, better change*—Use resources to demonstrate top management sponsorship, assign dedicated change managers, and empower employees to implement change.
4. *Right investment, right impact*—Understand which types of investments can offer the best returns, in terms of greater project success.

If you follow this or a similar model, then your company should be able to stay Lean over the long haul.

Next, we will investigate some ways to become even Leaner.

CHAPTER 13

Becoming Even Leaner: Partnering, Outsourcing (Private Labeling), Import (Logistics), Technology, (Lean) Six Sigma and Lean Systems

While many retailers and wholesalers have started on a Lean journey or at least done some continuous improvement, they need to realize that it's not just a "point" solution and that they need to look outside the "four walls" of their business for help through collaboration, technology, and ultimately making their own customized version of the Lean program.

While there is never a "one size fits all" in Lean, it is always instructive to look at cases and examples of how some of the concepts have been used, which we will do now.

Partnering

As discussed previously in this book, there are and have been many ways in which manufacturers or suppliers, third-party logistics services (3PLs), retailers,

and wholesalers collaborate with each other, including vendor-managed inventory (VMI), Quick Response (QR), Efficient Consumer Response (ECR), and Collaborative Planning, Forecasting and Replenishment (CPFR). Let's look at some examples to see how they were accomplished and the results as reported in the "Collaborative Planning, Forecasting and Replenishment in European Grocery Retail" report [Tenhiälä, 2003].

Condis–Henkel Example

Henkel is a German-based manufacturer of consumer goods and has been one of the CPFR pioneers in Europe. Henkel's CPFR pilot was run with a Catalonian retailer, Condis, during the year 2000. This project also included upstream supply chain partners, including Henkel's packaging suppliers, Cartisa, Suñer, and Reverplas.

Scope

In addition to Henkel's Spanish central warehouse, this project included two of Condis's distribution centers (DCs) and an undefined number of packaging suppliers.

The stock keeping units (SKUs) focused on were promotional items in the detergent category. Forecasts were aggregated at the Condis DC level. The forecast horizon was as long as eight weeks in order to provide usable information to the suppliers, and promotional planning was "frozen" (i.e., it couldn't change) for up to two months before the events. The forecast interval was reported as daily, with no point of sale (POS) data used, although DC on-hand inventory balances and shipments to stores were used.

Results

Some of the results reported from the pilot were:

- ▲ Sales forecast accuracy improved from 60 to 75 percent.
- ▲ DC service level (i.e., availability) was improved 1.5 percentage points to 99.5 percent.
- ▲ On-time deliveries were improved by 6 percent.
- ▲ Adjustments to the promotional plan were reduced from 68 percent to 25 percent.
- ▲ The collaboration enhanced the operations planning and inventory management at Henkel and its suppliers.
- ▲ All participants found the results to be satisfactory, and they identified new directions for further development and collaboration.

Veropoulos–Elgeka–P&G Hellas–Unilever Example

Hellaspar Veropoulos is the third-largest retail chain in Greece. Using an e-commerce platform, it started a CPFR-like project with three of its major suppliers: Procter & Gamble Hellas, Unilever Hellas, and Elgeka S.A. (a local representative for various international consumer products).

Scope

The 2001 pilot included five retail outlets of varying sizes. Five other outlets were used to give control information from the traditional "uncollaborative" operations.

The initial phase of the project covered a total of 120 SKUs from the three manufacturers, with the focus being on promotional and new product items.

Store-level forecasts were generated utilizing POS data. A single shared forecast was generated, which was automatically converted into an order forecast. This forecast was corrected by the manufacturers and then confirmed as an order.

Results

The pilot had two major impacts: (1) stockouts at the store level were reduced by up to 62 percent, and (2) in-store inventory levels were reduced by 9 percent. It was also reported that the benefits to the manufacturers performing direct store deliveries were larger than the benefits to those receiving deliveries through centralized DCs.

Liquor Control Board of Ontario Example

The Liquor Control Board of Ontario (LCBO) was established in 1927 to sell liquor, wine, and beer through a chain of retail stores. LCBO stores are generally the only stores allowed to sell distilled spirits in Ontario. The LCBO is one of the world's largest single purchasers of beverage alcohol products.

Scope

In 1999, the LCBO had embarked on a new retail strategy that included renovated stores, more dynamic product assortments, innovative merchandising programs and displays, and an enhanced shopping experience. It was so successful that it strained the board's supply chain because of the increased uncertainty of demand and unreliable supply. Its warehouses were pushed to capacity as on-time delivery rates fell and in-stock positions were weakened.

At the same time, the LCBO's supply chain was becoming increasingly complex, as consumer tastes were changing, resulting in a product portfolio expansion. In addition, it purchased alcohol products from 68 countries on 5 continents with long lead times.

The LCBO created a supply chain project team with the mission of building innovative supply chain solutions that would improve product flow across its network of partners, while collaborating efficiently and effectively.

The team members researched best practices and decided to use the VICS/GS1-Canada, (a member of GS1, the world's leading supply chain standards not-for-profit organization) CPFR process, which is a standardized, scalable, and repeatable approach to collaboration.

The LCBO developed and implemented a CPFR solution that was fully integrated, automated, and customized for the LCBO. The first pilot in 2003 involved 6 suppliers, and by 2005 the program had expanded to 21 suppliers representing 34 percent of the business.

The LCBO was able to provide suppliers with automated weekly transmissions that gave them visibility into sales, inventory, and open purchase orders.

Results

The benefits of the CPFR program could be seen in a variety of areas:

- *Promotional planning*—Suppliers developed promotional plans that had more impact.
- *Sales forecasts*—The program resulted in a more accurate forecast of promotional lift, and a better understanding of assumptions facilitated resolution and an improved understanding of the impact of promotions.
- *Order replenishment*—Operating efficiency improved along with better asset utilization.

As a result of the CPFR program, the LCBO now had shorter lead times, reduced safety stock levels, improved on-time delivery and order fill rates, and improved customer satisfaction [Miller, 2005].

Outsourcing (Private Labeling)

Private-label products or services are manufactured or provided by one company and sold under another company's brand; they are usually lower-cost alternatives to regional, national, or international brands. (Note: In recent years, some private-label brands have been positioned as premium brands as well.)

The general categories of private brands available to retailers are:

- *Premium*—Comparable or even superior to manufacturers' brands in quality with a modest price savings (e.g., Walmart's Sam's Choice).
- *Generic*—Targeted at a price-sensitive segment by offering a no-frills product at a discount price.
- *Copycat*—Imitates a manufacturer's brand in appearance and packaging. Generally perceived to be of lower quality and offered at a lower price (e.g., inexpensive fragrances that are promoted to be the "same as" the original).
- *Exclusive co-brands*—Developed by a national brand vendor and sold by the retailer. It is difficult for the consumer to compare prices for virtually the

same product (e.g., American Living by Polo Ralph Lauren at J. C. Penney and American Beauty by Estée Lauder at Kohls).

One indication of the growth of retail power has been that some companies have extended the concept of private label to identify a brand with a store, a concept known as the store brand (e.g., Walgreen's Nice!, Home Depot's Hampton Bay and Husky, and Target's Archer Farms). This can be more profitable than selling nationally advertised brands.

Other than some major retailers who operate manufacturing facilities, most retail private-label brands are outsourced to contract manufacturers, which come in a variety of forms, such as national and regional brand manufacturers with excess capacity and smaller manufacturers that primarily produce store brands.

In many cases, smaller companies have used private-label products for rapid growth. Smaller retailers usually don't have input into the recipes or packaging for the products they buy. Instead, for example, they purchase products from a specialty food company that uses its own recipes and then labels the products for the individual retail store.

Wholesalers also use private labeling to build their own brands. An example is C&S, the largest grocery wholesaler in the United States. Both the Best Yet and Piggly Wiggly brands are owned and licensed by C&S Wholesale Grocers.

Benefits

Private labeling can have many benefits to the retailer or wholesaler, such as:

- Greater control of production, marketing, distribution, and profits of the product or service.
- Cost and time savings as a result of a contract manufacturer's expertise in ingredient costs and trends.
- Partnering with private-label manufacturers that have R&D and marketing teams that can work with brand owners on packaging, design, and sales strategy.
- Creation of a personalized image that may lead to higher customer loyalty.
- Ability to take advantage of the general trend in customers' preference toward private-label products (while they were once considered a low-cost alternative, many store brands are now thought of as being on a par with name brands).

In Deloitte's 2013 American Pantry Study [Deloitte, 2013] of more than 4,000 participants, 88 percent said that they believe that private-label brands (sometimes known as generics) are just as good as their national brand counterparts, while 70 percent didn't feel that buying a private-label brand was a sacrifice and only 27 percent of respondents said that they planned on buying national brands more if the economy improved.

It appears that things have changed greatly since the early days of store generic products and their negative connotations, so that today, we have more widely accepted (and, in some cases, with "cachet") private-label brands.

So I think we can see that private labeling can be an integral part of a Lean (and growth) strategy for retail and wholesale organizations small and large.

Import (Logistics)

It's no secret that a large quantity of consumer and other durable items purchased today were manufactured overseas, as the trend toward outsourcing of manufacturing really took off in the 1980s in a variety of durable goods. From a Lean perspective, this is one way for retailers and wholesalers to lower costs (without denying the significant negative impact on businesses and employment in the United States).

As we've seen in recent times, there can be great risk when importing from overseas. Some examples include natural disasters (e.g., the earthquake and tsunami in Japan), fuel costs (the spike in fuel costs several years ago had a great impact on transportation costs, driving a recent trend in "near sourcing" to combat the issue), terrorism, political unrest, currency issues, protectionism, and so on.

In many cases, outsourcing has enabled retailers and wholesalers to offer lower prices on merchandise, but it has also significantly complicated our businesses from a supply chain and logistics standpoint (one of the topics covered in my previous book entitled *Lean Supply Chain and Logistics Management*). While I'm not going to delve too far into this topic here, we should look at how this longer, more complex supply chain, if managed improperly, can lead to inefficiencies and higher than necessary costs.

While manufacturers, importers and distributors, and large retailers typically import materials and finished goods directly from overseas companies, many wholesale distributors and small-to medium-sized retailers must go through other channels, such as overseas intermediaries, leading to delays, higher costs, and other inefficiencies.

In general, when dealing with global logistics, there are an extraordinary number of parties involved in the process. They can include customs agencies, freight forwarders, banks, ports, transportation companies (rail, truck, and ocean), and many others.

As a result, it is important to be aware of some of the potential issues and to use Lean tools such as value stream mapping to analyze and improve the processes.

Keys to Global Logistics Excellence

According to a 2006 *Supply Chain Digest* white paper entitled "The 10 Keys to Global Logistics Excellence" [*Supply Chain Digest*, 2006], there is a tendency to overestimate the savings in the first place, or perhaps the savings were lost because of poor execution.

The cost could be miscalculated because of a number of things, including higher than anticipated transportation costs or more buffer inventory required as a result of a longer replenishment time.

Poor execution can easily occur in the world of global logistics. The expertise required to handle the various languages, currencies, duties, tariffs, and other such factors can be overwhelming, even for someone who has a lot of experience in this area. Lack of appropriate technology can also be a detriment in this area.

Most if not all of the 10 keys to global logistics excellence mentioned in the white paper just cited pertain to Lean and should be considered when looking for waste in this area. They are:

1. *Total delivered cost management*—There is a plethora of logistics costs when companies are sourcing and shipping globally. They include domestic and international transportation and distribution costs by product and route. You also need to include duties, tariffs, and other customs costs. Having automated systems to track and measure performance of this information helps, of course.
2. *Global logistics process automation*—The more automation the better, especially in terms of visibility. Also, a lack of automation results in too much time being spent on manual activities instead of planning. Global logistics execution is complex, with as many as 25 or more hand-offs in one shipment with multiple parties. Many enterprise resource planning (ERP) and logistics systems don't support global logistics as well as the users require.
3. *End-to-end visibility*—Global visibility and event notification are critical to minimizing waste in your supply chain. This is useful not only for tracking but also for spotting and reacting efficiently to exceptions. Timeliness and accuracy are extremely important when managing your global supply chain. Examples include electronic data interchange (EDI) with ocean carriers and Web portals to 3PL providers and customs brokers. This allows companies to make the most of their internal labor and other global overhead resources while being in control of their outsourced global supply chain.
4. *Supplier portals and advance ship notices (ASN) capabilities*—Even though you may have adequate transportation visibility, you may still be lacking the status of each order at the foreign factory or the details of what's in each container. That's where ASNs and supplier portals help you to get early notification of this type of information, thus giving the incoming DC advance notice and helping to plan inventory better (and thus reduce waste). Many companies still receive this information via fax and in many cases can't be 100 percent sure of its accuracy until they open the container.
5. *Total product identification and regulatory compliance*—Security, import/export restrictions, customs, and other safety and regulatory requirements hinder the continuous flow of the global supply chain, both slowing and adding cost to it. Technology can be of some assistance in this area, such as the use of radio-frequency identification (RFID) systems and software that works with denied screening and other regulatory requirements.
6. *Dynamic routing*—Even on repeated routes, to reduce cost and increase agility, shippers are now looking for different combinations of carriers, routes, and freight forwarders similar to those they have used domestically for years. Not

only do they get lower rates, but they can have a more agile supply chain with more dynamic alternatives (such as direct ship, drop ship, cross-dock, and so on). This can also lessen the risk of events such as the recent earthquake and tsunami in Japan, where it would have been ideal to have a backup supplier of semiconductors with all the logistics costs and processes known in advance.

7. *Variability management*—As we know, variability both is a contributor to waste and has a huge impact on both inventory and customer service, and it's even more prevalent in the area of International delivery times. By using performance measurements and tools such as root cause analysis, supply chain managers can attempt to minimize the variability for better, more predictable global supply chain and logistics performance.
8. *Integrated international and domestic workflow*—Until recently, there really weren't transportation management systems (TMSs) that integrated domestic and international logistics, almost forcing companies to look at them as two separate shipments. There is still a lot of improvement to be made on this front, but it appears that there should be further progress in the next two or more years. This will also allow for more centralized international logistics, similar to what has existed on the domestic side for years.
9. *Integrated planning and execution flow*—The idea here is to attempt to have technology on the global logistics side that follows the planning and execution process (i.e., automated and integrated). Ideally, this should be real-time or close to it, and, as was pointed out earlier, much of the international freight movement data is still collected manually. This is truly the ultimate way to make sure your global supply chain and logistics function is Lean.
10. *Financial supply chain management*—It is especially critical on the global logistics side that financial processes for things like letters of credit, financial settlements, and other import/export documentation don't interfere with the flow of goods. So it is important to have internal personnel as well as partners and software vendors who are skilled at this to make sure that international transportation planning and execution are integrated and as efficient as possible.

Technology

As you can see from this discussion of retailers and wholesalers partnering with customers and suppliers, collaboration can't be done efficiently without the aid of technology. Luckily, there is an abundance of relatively low-cost technology that can connect suppliers, manufacturers, 3PLs, wholesalers, and retailers via the Internet or EDI.

Technology in Partnering

What follows are some examples of manufacturers, retailers, and wholesalers partnering with each other using technology to become Leaner, while also using it for a competitive advantage.

Black & Decker, Home Depot, and Lowe's Example

For a retailer or wholesaler, it is important to see how manufacturers view collaborative partnerships like CPFR to see the "big picture" of how it works and what benefits everyone can achieve.

To better support its existing alliances with two superstore retailers—Home Depot and Lowe's—supply chain leaders at Black & Decker Hardware and Home Improvement (HHI), a manufacturer and marketer of building products for the residential and commercial markets, sought to establish one synchronized view of demand throughout its supply chain.

Upon completion of the project, the firm was able to implement a CPFR strategy backed by enabling technologies and an aligned business information systems (IS) team. This allowed the manufacturer to realize its large retail partners' requirements of greater than 98 percent fill rate *and* on-time delivery. It also realized benefits beyond improved collaboration at retail to help reduce cycle time to meet homebuilders' make-to-order configured product requirements within 14 days.

To reach this goal, Black & Decker HHI created dedicated demand forecasting teams to work specifically with personnel employed by Home Depot and Lowe's. As the organization had no central planning software in use at the start, CPFR was a labor-intensive process; planners massaged large amounts of product data in spreadsheets received from retailers and looked at historical sales to project demand based on an analysis of trending and seasonality.

In order to gain better visibility of its supply chain and do it in a more efficient manner, Black & Decker HHI developed three software implementations to meet the requirements of the various planning groups.

Black & Decker HHI started in its area of greatest need: supply chain planning. After holding a functionality and software review, the company selected JDA Demand from JDA Software Group.

The solution was configured to integrate POS data from Home Depot and Lowe's, creating one single process for its line reviews, product promotions, introductions, and price changes. The participants were able to compare forecasts and shipment history as well as POS data and order history to improve their forecasts.

Black & Decker HHI then implemented JDA Master Planning and then, soon after, added JDA Fulfillment to the technology mix to completely synchronize supply and demand.

Now, with full visibility into its demand and supply chain operations, Black & Decker HHI has built truly collaborative relationships with its retail customers.

These results, in addition to a 10.4 percent improvement in forecast accuracy (helping to reach the retailers' goals of more than 98 percent order fill and on-time delivery), have prepared Black & Decker HHI to support future retail relationships [Ackerman and Padilla, 2009].

HDA, Lowe's, Dollar General, and Michaels Example

HDA is a U.S.-based distributor of books and magazines to the home improvement, craft, home décor, discount, kids', office, and cooking specialty retail channels. It is a "turnkey" supplier for retail partners such as Lowe's, Dollar General, and Michaels.

> The Company provides sophisticated supply chain management services based on Category Managed Merchandising (CMM) and VMI models. . . .
>
> In a VMI/CMM process, HDA has full responsibility for selecting the best product and mix, creating the plan-o-gram for the section, initiating the orders, accurate order fulfillment, timely shipping, reliable store level service, and maintaining all metrics including sales, inventory turn, margin, and ROI. This is accomplished through EDI communication of store level sales and returns, analysis of industry movement data available at the individual title level and periodic store-level input from field service personnel to HDA's VMI software system.
>
> The VMI software system generates the optimal order, based on agreed to or set service parameters. The system then integrates with the Company's logistics systems, including the Perfect Pick and Pick-to-Light Systems both of which use RF scanners, to process orders with a fill rate accuracy of 99.92 percent and ship them with an on-time rate of 99.98 percent. Upon shipment, the company is then invoiced electronically as well.
>
> . . . HDA's retail clients typically use 20 to 30 percent less inventory on average in VMI stores versus traditional supply chain partnerships. Out-of-stocks and missed opportunities are kept to a minimum through HDA's systems. In addition, sales are typically 10 to 15 percent higher in VMI stores. [HDA, 2013]

Technology in Retail Marketing

Besides enabling partnering in operations, technology has also started to revolutionize retail marketing. Retail marketers now have access to vast amounts of data to find, target, and retain customers. There is now technology available to perform the analytics needed to convert these data to real, usable information.

Kelly Kennedy, senior vice president of enterprise sales at Infogroup Targeting Solutions [Kennedy, 2013], finds that there are five big data trends that are transforming retail marketing. They are:

1. *Growing cross-channel data volumes*—Mobile technology, tablets, and social media have increased the growth of available customer data. Now retailers have not only the basic demographic information about a customer, but purchase history, call center interaction, mobile/social interaction, supply chain data, and more.
2. *Increasing investment in technology*—Storage is cheap, and it's leveling the playing field for many companies when it comes to "big data," allowing for improved analytics to provide more insight into their customers.

   In 2013, retailers will spend nearly $2 billion on business intelligence and $9.4 billion on infrastructure. Macy's attributes a 10 percent increase in store sales to improved analytics capabilities.
3. *Solving the multichannel puzzle with data*—Retailers are analyzing a huge amount of customer-behavior data to understand how customers are researching and buying products. Insights reached through analyzing transaction data, foot traffic, and in-store checkout wait times have led to shifts in marketing strategies and in-store tactics using in-store kiosks, free Wi-Fi, and sales staff members with mobile devices that allow them to better serve Web-savvy customers on the spot.

   Walgreens found that customers who shop both in-store and online spend 3.5 times as much as customers who favor only one channel. So it's also important not to ignore one channel in favor of the other.
4. *Improving personalization*—Big data allow retailers to adapt their communication and sales techniques to life events and preferences. Research in a *Harvard Business Review* blog found that personalization can deliver five to eight times the ROI on marketing investment and boost sales 10 percent. Consumers are OK with sharing personal details if doing so offers them some kind of reward.

   An emerging new technology known as "next best offer" (NBO) offers the convergence of real-time data analysis and mobile offers by reaching consumers at the right time, in the right place, through the right channel in a personalized manner.
5. *Segmenting the most valuable customers*—The goal is to determine who are the most profitable customers. Technology allows retailers to analyze the behavior and needs that drive individual customers, resulting in timely targeted offers.

This use of technology to analyze huge amounts of customer data can certainly make the sales and marketing function Leaner and more efficient, while at the same time increasing revenues.

E-commerce

Of course, e-commerce is the ultimate use of technology to create and expand a retail business (and, from a Lean perspective, can be a great alternative or supplement to brick-and-mortar to reduce operating and other costs).

So it's a good idea to think about where to look to "Lean out" these technology-based e-tailers, as we have found that in many cases, overspending and overexpansion, rather than monitoring cash flow and focusing on profitability, have sunk many e-tailers early on.

Therefore, what can an e-tailer do from a Lean perspective to help ensure success? Here are some thoughts on the subject:

- *Substantial Gross Margins*—Operating on a negative gross margin is a recipe for disaster. "Many of the failed e-tailers of 2000 established business plans that were intended to operate on negative gross margins for an extended period of time, a fundamentally flawed business strategy," Yankee online retail analyst Paul Ritter told the E-Commerce Times. . . .
- *Customer Acquisition*—The Yankee report predicts that e-tailers who spend in excess of $20 to $40 for each paying customer, or 50 percent or more of revenues on sales and marketing, will have a difficult time surviving. . . .
- *High Conversion Rates*—"Rarely do online retailers earn enough gross margin to cover acquisition costs with a customer's first purchase." . . . Yankee estimated that the average conversion rate for online retailers is approximately 1 percent. . . .
- *Value Proposition*—Ritter noted that to generate sufficient customer loyalty, a company must have a potent value proposition. . . .
- *Customer Friendly*—"Customers must be able to find the products and information they are looking for quickly and with a minimal amount of effort and clicks," Ritter said. . . .
- *Effective Product Fulfillment*—Fulfillment costs have created major cash flow headaches, especially for companies such as Amazon.com and Webvan, according to Yankee. . . . Yankee predicts that there will be a significant trend in online retailers outsourcing their fulfillment to third parties, who have already built the infrastructure and have established a core competency in this area. An efficient process for managing returns is also critical, the report said [Mahoney, 2001].

M-commerce

As e-tail customers move from e-commerce to m-commerce, using smartphones and tablets to shop, the question of Lean becomes, how can you make the site operate smooth and fast? We're talking lead time here, which is the time it takes for a Web page to travel from a computer server to a smartphone.

One of the best at this is Office Depot, ranked 53rd in the Internet Retailer Mobile Commerce top 300 rankings. Its m-commerce site home page had the quickest page load time and a 99.91 percent success rate, which is the percentage of the time that a page loads completely and successfully, making it the highest of 30 retailers on the index in those categories. Another successful e-tailer has been Barnes & Noble, the book retailer and e-tailer, which has been ranked among the top three retailer sites every week on the 30-retailer Keynote index since November 2012 [Siwicki, 2012].

A Lean and efficient m-commerce site is especially critical, as it is predicted that by 2015, more U.S. Internet users will access the Web through mobile devices than through PCs, and sales of smartphones will continue to exceed those of all other types of computing devices combined (including tablets), according to Igor Faletski, the CEO of Mobify [Faletski, 2012].

Additional critical facts that Faletski pointed out were:

1. *Customers spend more time on their mobile devices than on their desktops*—Tablets have emerged as the third digital screen in consumers' lives, in addition to desktops and smartphones, and have become a lucrative customer segment for retailers.
2. *Mobile shoppers are more focused*—Mobile searchers and shoppers are very task-focused, are specific in what they're seeking, and want information or assistance to help them make buying decisions at the point of sale (88 percent of consumers searching on a mobile device will make a purchase within 24 hours). They want to research products and make a purchase quickly. So as a result, there needs to be minimal, but to the point, content on mobile commerce websites.
3. *Click-through rates are higher on mobile than on desktops*—Smartphones and tablets are showing higher click-through rates for search advertising. A Marin Software study determined that consumers are more likely to click on search results ads when using smartphones or tablets than when using a desktop or laptop.
4. *Mobile shopping peaks at night*—Social media activities and e-mail tend to be used during commuting time, but not online shopping. Tablet usage spikes at night, and people searching on their mobile phones overlaps both periods, but grows steadily during the day and peaks in the early evening. So this should be considered in terms of Web site capabilities and staffing.
5. *The importance of the mobile Web*—Although we are hearing a lot about apps lately, mobile Web sites (i.e., sites that are reached by browsing, not through an app) tend to get somewhat overlooked. While apps can be part of a mobile strategy, and are usually for repeat customers, it's important to start with your mobile Web site and then supplement your mobile strategy with apps.

(Lean) Six Sigma

6σ

Six Sigma attempts to improve the quality of process outputs by identifying and removing the causes of defects (errors) and minimizing variability in manufacturing and business processes. It uses a set of quality management methods, including many statistical methods, and creates a special infrastructure of people within an organization (Champions, Black Belts, Green Belts, Yellow Belts, and so on) who are experts in these methods.

An individual Six Sigma project carried out within an organization follows a defined sequence of steps and has quantified value targets—for example, process cycle time reduction, customer satisfaction, reduction in pollution, cost reduction, and/or profit increases.

Six Sigma can be thought of as being complementary to Lean, as Lean looks at an entire value stream from customer to supplier and Six Sigma tends to look at an individual process's variability. As a result, there is now "Lean Six Sigma."

A Lean Six Sigma project includes Lean's waste elimination philosophy and tools, plus Six Sigma concepts based on removing defects and minimizing variability to improve quality. The DMAIC (Define, Measure, Analyze, Improve, and Control) tool kit of Lean Six Sigma comprises all the Lean and Six Sigma tools. The training for Lean Six Sigma is provided through a belt-based training system similar to that of Six Sigma.

According to a Bain & Company 2008 brief entitled "Lean Six Sigma for the Services Industry," we know that, although:

> Lean Six Sigma was originally devised to eliminate waste and improve manufacturing quality to no more than 3.4 defects per million opportunities . . . the method—made popular at companies like General Electric Co., Xerox Corp., and Johnson & Johnson—is increasingly finding a home in the services industry.
>
> We have seen banks use Lean Six Sigma to support their growth strategy; financial services companies to put mergers back on track; energy companies to lower costs; telecommunications companies to improve customer service; and retailers to increase efficiency while boosting customer service in the store. But Lean Six Sigma's growing popularity in the services industry masks a downside. Many organizations have trained and deployed legions of Lean Six Sigma experts—known as black belts—only to see little value result from their work. In a recent Bain & Company management survey of 184 companies, 80 percent say their Lean Six Sigma efforts are failing to drive the anticipated value, and 74 percent say they are not gaining the expected competitive edge because they haven't achieved their savings targets.
>
> Drilling deeper, we discovered that mobilizing large and costly squads of black belts in some cases actually slows performance improvement efforts. Managers are unsure how best to deploy the Lean Six Sigma experts and too often black belts treat all problems, big and small, with the same approach, resulting in less-effective solutions. Moreover, they fail to prioritize the improvements that will make the biggest difference. [Guarraia et al., 2008]

Therefore, retailers and wholesalers, at least in my opinion, are better off using Six Sigma and its statistical tools and infrastructure as a complement to Lean, which is more of a companywide tool.

That being said, Bain & Company points out that in its opinion, the biggest gains from Lean Six Sigma occur when using some kind of diagnostics tool such as its "x ray," which helps to identify opportunities. Bain's diagnostic x ray, which is done by trained Black Belts, consists of three fairly basic steps:

1. *Enterprise value stream map*—To identify primary candidate areas to reduce cost by reducing wasted time and materials.
2. *Benchmark*—Identify shortcomings and set improvement targets.
3. *Prioritize*—Determine which process improvements will generate the greatest results.

Once the team has gone through this process, it can use the Lean Six Sigma DMAIC process of Define, Measure, Analyze, Improve, and Control on these areas.

Here is an example of the use of this diagnostic tool at one of Bain's retail clients.

Big-Box U.S. Retail Leader Example

A large U.S.-based retail market leader needed to stem competition from general merchandisers.

The Bain & Company x-ray team used value stream mapping for nearly 100 routine tasks. This process included documenting activities such as opening and closing stores, pricing, taking inventory, restocking shelves, and filling orders placed through the retailer's Web site.

The team then prioritized each task based upon the activity's perceived importance.

Among other things, the team looked at:

- A special form for when a customer requested price matching.
- The system for filling online orders.
- The system used to keep a steady stream of high-profit items on store shelves.

The team found $50 million in annual cost savings from some relatively small process improvements.

Its top recommendations included the following:

- *Redesign the restocking calendar*—Employees needed to more easily see which products to check at what times each day throughout a month (estimated annual savings: $10 million).
- *Eliminate time-consuming customer forms*—This saved an estimated $3 million annually ($2.3 million from labor and $800,000 from paper).
- *Improve the online order pickup process*—The team recommended reorganizing material receiving and storage as well as communication of orders received. The retailer would save another $1.2 million from this improvement.
- *Standardize the daily setup of cash register drawers*—The retailer needed to analyze how much change was needed for a shift ($700,000 annual savings).

When the retailer finally unleashed its Black Belts, the diagnostic allowed it to focus immediately on creating results that matter [Guarraia et al., 2008].

Staples Example

Another example of Lean Six Sigma used by the office supply retailer Staples with the help of the consultant Accenture accomplished quite a bit.

> Accenture's help with The Staples Lean Six Sigma program has been the impetus for dozens of improvements that have generated tens of millions of dollars in benefit for Staples and produced a 10-fold return on the company's investment in the process improvement program:
>
> - *Rebalance lease negotiations and improve architectural and construction processes*—Efforts that shaved four weeks off the time needed to open a new store, leading to increase in sales equivalent to eight new stores annual sales.
> - *Streamline the item-order cycle and have promotional items arrive at stores closer to sale dates*—Efforts that freed space (especially in smaller stores) and generated inventory savings of $3.3 million.
> - *Reconfigure the loading dock layout, eliminate extra handling of merchandise, and establish a "receiving and put away team" within one fulfillment center*—Efforts that improved on-time to due date performance by 21 percent.
> - Consolidate freight moving from suppliers to Staples' distribution and fulfillment centers, which achieved 50 percent of the budget reduction stretch goal for the year.
>
> [GoLeanSixSigma.com, 2013]

Additional Lean Six Sigma Retail and Wholesale Examples

While these are great examples of Lean Six Sigma, small-to medium-sized retailers and wholesalers often don't have the resources to send out teams of Black Belts. However, they can give some training and support to these activities as part of a larger Lean program.

What follows are some more discrete retail examples that also apply to wholesale and were described in an Accenture white paper entitled "Solving Retail Problems Using Lean Six Sigma" [Curtis et al., 2008]:

Example 1: Reduction of Delivery Errors

Problem—A large U.S. retailer was incorrectly loading more than 20,000 boxes of customer orders per year, putting boxes on the wrong delivery trucks. These errors increased labor and transportation costs to redeliver material and reduced customer satisfaction.

Solution—The project team identified and implemented several solutions, including permanent route changes, installing large whiteboards with planning data, and moving the label printing to a more visible location. These relatively easy and low-risk solutions resulted in $30,000 of cost savings and a 65 percent reduction in errors for the DC, and if replicated at other centers had the potential of another $200,000 of cost savings.

Example 2: Reduction of Damaged Inventory

Problem—Damaged inventory levels in a leading global retailer's U.S. stores averaged $11.5 million. This needlessly tied up working capital and slowed down replenishment of inventory in stores.

Solution—The team found that 30 percent of the reason for return codes entered were incorrect. By using tools such as Lean value stream mapping, process balancing, and statistical analysis, the team discovered several critical root causes: (1) process failure points were uncovered before the damaged product inventory was placed in the bin, (2) cashiers were not properly trained on returns and exchanges, and (3) there was no audit process for damaged product reports.

A cross-functional team of corporate and store inventory employees decided to redistribute workload and accountability between the store inventory and service desk associates to increase inventory accuracy. The project resulted in inventory reductions of 26 percent or $3.5 million, and the cycle time of items in the damaged inventory was reduced by 37 percent to less than 10 days.

Example 3: Improved Freight Consolidation

Problem—A large global retailer incurred higher than necessary freight expenses by missing opportunities to consolidate freight transported from its suppliers to its DCs.

Solution—The retailer had a freight optimization process that was executed twice each business day, but if it could consolidate additional supplier shipments, it could potentially reduce annual inbound freight costs by more than $750,000.

The team found reasons for inefficient freight optimization, which were: (1) suppliers had to request routing instructions at least 48 hours in advance of their ship date, (2) suppliers were given instant routing responses unnecessarily, and (3) the current allocation of resources did not provide freight optimization process support. All of this reduced shipment visibility and the time available for the retailer to analyze opportunities and consolidate freight.

By removing suppliers' ability to receive instant routing and reallocating internal resources to freight optimization, the team was able to eliminate four defects in the process and save $752,000.

A Lean System

From the early stages of Lean (before it was even known as *Lean*), companies realized that it is important to make a program their own. A good early example of that is the Toyota Production System (or TPS), as shown in Fig. 13.1.

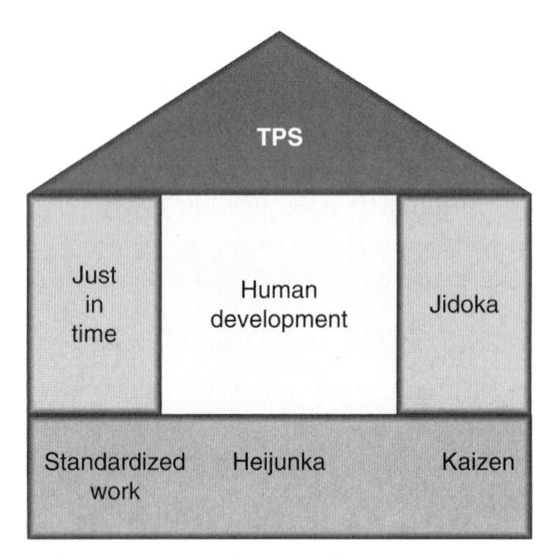

Figure 13.1 Toyota "House of lean."

Toyota Production System

The TPS is both a philosophy and a system that organizes manufacturing and logistics for the automobile manufacturer and includes interaction with suppliers and customers.

Some of the concepts that are central to the TPS are:

- *Assembly components*—Components are placed in a cab for easy access rather than on shelves adjacent to the assembly line.
- *Respect for people*—Employees are treated as knowledge workers and are empowered to stop production, come up with ideas, participate in quality circles, and so on.
- *Just-in-time (JIT)*—Parts and supplies are delivered just as needed in the quantity needed.
- *Level schedules (heijunka)*—Models are mixed on production lines to meet customer orders by coordinated sequencing of very small production batches.
- *Jidoka*—Machines have built-in devices for monitoring performance and making judgments.
- *Pull system*—Units are produced only when more production is needed, using a kanban system.
- *Standard work practices*—There are rigorous, documented procedures for production (and elsewhere).
- *Minimal machines*—Proprietary machines are made specifically for Toyota applications.
- *Kaizen area*—This is an area where suggestions are tested and evaluated.

As you can see, some of the concepts are general in Lean thinking today, while others are specific to both the auto industry in general and Toyota specifically.

I've done a fair amount of Lean training and consulting, and I have found that the companies that have implemented Lean successfully have done it in a similar manner—combining fairly standard Lean concepts and tools, but making sure that they fit the industry and the company in terms of their culture and processes.

A more recent example of an organization creating its own customized Lean system is the DuPont Production System (DPS), which DuPont has employed at its own company and also uses on a consultative basis to assist clients.

DuPont Production System

According to the company's Web site [DuPont, 2013]:

> The DuPont Production System is a holistic model for management that engages all employees with a common set of tools and techniques. Based on DuPont best practices, this system focuses on eliminating inefficiencies and waste, solving problems, and measuring progress. We develop worker and corporate core values, strengthen leadership and organizational structures, and implement focused processes and actions that establish this system throughout the organization. This system has helped DuPont and our clients to develop a standardized approach to production, defining realistic efficiency goals with attainable, measurable metrics, and targets for financial performance and beyond. . . . This unique system is designed to maximize return on investment and net present value. Rapid gains in performance help to fund future transformation—creating a sustainable investment. The DuPont Production System reliably and comprehensively helps organizations reap financial benefits while establishing a sustainable culture of operational excellence.

This statement alone sets the tone for a Lean (or at least continuous improvement) culture at DuPont. The company has also established "Centers of Competencies" to identify, develop, share, and support best practices throughout the organization.

Furthermore, in a manner similar to the TPS "House of Lean," it has established "seven elements of DuPont asset productivity" to improve and sustain operational efficiency, productivity, and competitiveness. They are:

1. *Maintenance and reliability systems*—To help find and resolve the root causes of process unreliability and poor equipment performance in order to lower maintenance costs.
2. *Manufacturing capacity processes*—To increase throughput and reduce cycle time by eliminating bottlenecks and enhancing workflows.
3. *Energy optimization*—To enhance energy efficiency, safety, reliability, and operability through technology.
4. *Facilities infrastructure processes*—To use an audit process to help minimize maintenance costs at facilities.

5. *Mechanical integrity*—To enhance regulatory compliance and, at the same time, reduce emissions and operational failures.
6. *Product quality and process controls*—To use DuPont tools such as dynamic analysis, control strategy synthesis, design for process operability, application of statistical methodologies, and quality improvement practices.
7. *Organizational capacity improvements*—To implement a change management system, so that all employees have the proper skills and appropriate tools to maximize asset efficiency and life [DuPont, 2013].

The "deliverables" of the DPS, which are tied to these elements, are:

- Integrated strategies and operational plans
- Productivity and asset effectiveness along supply chains
- Technology ownership and integration along supply chains
- Advancing core values
- Capability building—people and organizational development
- Mindsets and behaviors that foster engagement and superior execution

Now that we've explored various ways in which retailers and wholesalers can implement a Lean program throughout their organizations, both strategically and operationally, we will explore ways to get the most out of your employees in order to create longer-lasting, sustainable results.

PART III

Lean Forward

CHAPTER 14

Critical Thinking and Continuous Improvement: Methodology, Education, Training, and Analytics

As was mentioned at the end of Chap. 13, every organization will want to put its own stamp on its Lean program or system, but at the same time, the company has to have a specific implementation methodology or approach in mind if it is to have a "road map" to success.

Methodology or Approach

I have one approach that has worked for me and my clients in the past and that at the least can be a good starting point for discussion purposes (there are many to choose from). I call it the "4i" approach (see Fig. 14.1). The steps are:

- *Instruct*—Educate the entire organization about Lean principles, concepts, and tools.
- *Innovate*—Use tools such as Lean assessments and value stream mapping (VSM) to document current-state processes, then determine future-state processes with a goal of reducing non-value-added activities.
- *Implement*—Identify specific projects through assessments and the VSM process to reduce waste, such as 5S (workplace organization), setup and batch size reduction, work cells, pull/kanban, and total productive maintenance (TPM).
- *Improve (continuously)*—Engage in cultural, job, and process changes to encourage long-term success through implementing Lean concepts at your organization (in some ways, this is both the beginning *and* the end of this particular methodology).

No matter how you decide to roll out a Lean transformation program in your organization, it needs to include education of the workforce (instruct), the free

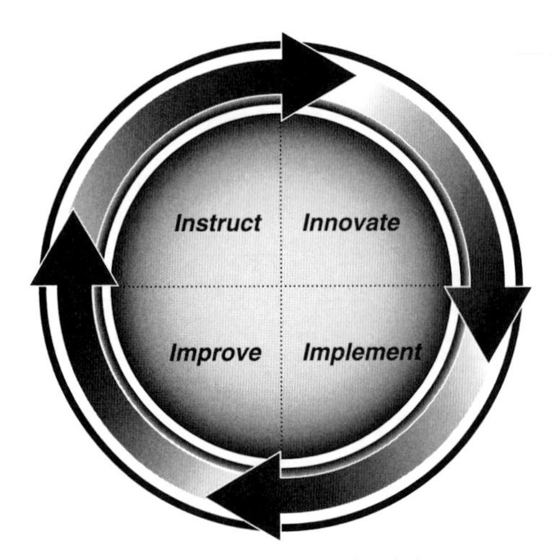

Figure 14-1 4i's methodology.

use of critical thinking (innovate), a schedule of kaizen events with sound project management practices and structure (implement), and a supporting culture for continuous improvement.

McKinsey & Company [Fine et al., 2008] suggests using a balanced approach to successfully implement Lean and Six Sigma in the organization. It points out that often the "softer side" (i.e., developing leaders and linking the boardroom with the shop floor) is neglected in favor of the "harder side" (i.e., operational tools and approaches). This is often the case, as implementing Lean's softer side can be hard because it forces all employees to commit themselves to new ways of thinking and working.

McKinsey's balanced approach looks at the design and implementation process for continuous improvement as having three overlapping aspects:

1. *Technical (operating) system*—How corporate resources are deployed to meet customer needs at the lowest cost; this includes various operations, logistics, quality and maintenance processes and systems, quality systems, and labor allocation.
2. *Management infrastructure*—The structures and processes that the organization uses to manage its processes and systems to meet company objectives, including performance and talent management as well as organizational design, roles, and responsibilities.
3. *Mindsets and capabilities*—This refers to the way employees think about their work and their workplace, including leadership, direction, collaboration, accountability, trust, skills, and ultimately continuous improvement.

Organizations that are successful in integrating a program's soft and hard elements while developing their line managers' Lean leadership skills can gain larger, more sustainable benefits. An example of this follows.

Example: North American Distributor's Approach

In the McKinsey article, the authors point out an example of a North American distributor that was facing greater customer requirements with narrowing profit margins throughout its network of 70 distribution centers (DCs).

Management used this more balanced approach, looking not only at technical changes, but also at structure and process, including how the changes would affect employees.

Team leaders determined that labor balancing would be an important improvement, so they combined customer order picking with truck loading. The idea was to increase productivity by balancing labor with demand.

As they were using a balanced approach, they also changed the performance management system to reflect the new work methods. In the past, pickers had been measured quantitatively and packers either qualitatively or not at all.

Creating a team-based system to balance the speed and accuracy of picking and loading also helped encourage employees to collaborate and have a common goal. Furthermore, the leaders created a visual tracking system to reinforce the new behavior that displayed shifting workloads that they needed to attend to.

Management also had to address the mindsets of workers, as many of them were longtime employees. There was also a bit of a rivalry between pickers and loaders. With this in mind, the company provided some interpersonal skills training for the supervisors in order to deal with this somewhat touchy situation.

The net results were impressive, as within six months the participating DCs were 10 to 15 percent more productive, on-time deliveries had improved by 5 to 10 percent, and reported shipping errors were down by 33 percent. Worker job satisfaction had actually risen by 10 percent as well. Later analysis revealed that half of the productivity gains were a result of the software elements previously described.

Example: Spanish Supermarket Chain's Approach

In another example, as pointed out in the Harvard Business School newsletter *Working Knowledge* [Hanna, 2010], the Spanish supermarket chain Mercadona focuses on the "last 10 yards" of the supply chain, which lie between the stores' receiving dock and the customer's hands.

It does this by trying to make manufacturing principles similar to the Toyota Production System work in retail while offering low prices and personalized customer service. Mercadona also pays its employees above-average wages and offers them four weeks of training, compared with one week in the United States.

Mercadona cross-trains employees so that productivity isn't directly tied to store traffic, but instead is tied to efficiency and customer service. Employees are also given their schedules a month ahead and are always on the same shift, resulting in a very low turnover rate of 3.8 percent. This stability, which is not often found in retail, gives employees a feeling of stability.

This philosophy contributes to sales per employee that are 18 percent greater than those of competitors (and 50 percent more than those of U.S. supermarkets) and sales per square foot that are double the average for U.S. supermarkets.

Mercadona has also taken the opposite approach to that used by most of its competitors, offering 43 percent fewer products per square foot of retail space. It is successful in this by having field employees that the firm calls "prescription instructors" visit stores in their area and speak with customers. This feedback has resulted in improvements to product, the supply chain, and employee productivity in many cases.

Mercadona chooses to invest in store processes and store labor to improve operations and focuses more on long-term profitability and the impact its decisions have on its customers, employees, suppliers, and society than on short-term profits.

I think it's pretty clear at this point that training is a critical part of the success of any continuous improvement program, so let's look at the various methods that are used today.

Training

Most of the Lean training and implementations that I've facilitated over the years have been delivered primarily by using the *train-do* method. This involves starting with some initial basic training in Lean concepts and tools, typically followed by team-based "critical thinking" using processes such as VSM and brainstorming sessions to come up with areas for improvement. The recommended improvements are then implemented by the team through scheduled kaizen events, using Lean concepts and tools.

This type of training combined with my university experience as an adjunct professor has enabled me to come to some conclusions concerning what works and what doesn't work.

There are certain training methods and tools that can be successfully applied to Lean, including its application in the supply chain and logistics function.

Training Methods

Traditional Methods

Raymond A. Noe, in his book *Employee Training and Development*, published by McGraw-Hill in 2002, points out that traditionally there have been three major methods of training: (1) presentation, (2) hands-on, and (3) group building. The train-do method typically combines all three methods to some extent.

Presentation Method

In the presentation method, the group is passive and primarily listens to information presented to it through lectures supplemented by audiovisual means (e.g., slides and videos). This is typically one-way, from the trainer to the audience. It is a relatively inexpensive, efficient way to transfer knowledge to a large group.

This method can have variations, such as team teaching, guest speakers, and panels, to make it a bit more interactive.

Hands-On Methods

Hands-on methods usually require the trainee to be actively involved in the learning process. These methods can take the form of on-the-job training (OJT), simulations, case studies, business games, role playing, and behavior modeling.

OJT can be used for new or inexperienced employees who need to get up to speed on a new job, such as operating a piece of equipment, or for cross-training purposes. This can be taught by a current employee, by apprenticeship, or through self-directed learning. These days, you don't see as many apprenticeship programs as in the past. The most common reason given seems to be that "it's hard to find good people to develop." However, they can still be found in skilled trades and are typically sponsored by the company or the union.

To be effective, the trainer should have lesson plans, checklists, procedure and training manuals, and progress report forms available at the time of training.

In this age of the Internet and intranets, self-directed learning can be advantageous, as the trainees can learn at their own pace, learn on their own schedule, and require less supervision. However, the trainees must have a good amount of self-motivation to complete the course.

Simulation Games

Simulations are quite useful in that they mimic a real-life situation, and the decisions made by the employee have outcomes similar to those that would occur in the workplace. Simulations can be useful in developing teamwork, production, process, and management skills. In fact, in 2009, I developed a Lean supply chain and logistics management training simulation game for this specific purpose that is useful for manufacturers, distributors, and retailers (http://www.enna.com/lean_supplychain/).

Case Studies

Case studies are useful in showing trainees things that may happen in the workplace and allowing them to develop and use critical thinking skills as well as developing teamwork skills. They are a useful tool not only in the university setting, but also in the workplace, as they allow the trainees, while guided by a facilitator, to see how other companies deal with business issues like theirs and then transfer some of that thinking to their own workplace.

Role Playing

In role playing, trainees act out characters that are assigned to them in a specific scenario. The idea is to focus on the responses of participants and how to deal with different situations. This can be especially useful and fun when doing simulation training games.

Behavioral Modeling

Finally (and probably least used), there is behavioral modeling, where trainees are presented with a model with specific behaviors that they attempt to repeat. This works better for learning skills and behavior than for learning information or facts.

Group Building Methods

Group building methods are used to improve team or group effectiveness. The trainees share ideas and experiences, build a group identity, and grow to understand their teammates' strengths and weaknesses.

There are a number of group building methods. There are more adventure-type methods such as Outward Bound, where the team goes through wilderness or outdoor training to develop teamwork and leadership skills. These are especially effective for improving problem-solving and conflict management skills. The key to this type of training is a wrap-up at the end where the results and how to apply them in the workplace are discussed.

Team Training

Team training, discussed in Chap. 10, involves getting the team members to work together to reach a common goal. The optimal result of teamwork is that the team members learn to identify and resolve issues together. In order to be successful, the team members must be properly trained and supported by management.

Action Learning

In action learning, the team or group works together on a real issue or problem and creates an action plan to resolve it. This type of group method can be used to make changes to processes, improve the use of technology, or improve customer satisfaction, for example.

Individual Learning

I'd like to add to the list individual learning using technology, as it is becoming more common in retail these days.

A leader in this method is the Hard Rock Cafe restaurant chain. Hard Rock supplements some of the traditional methods discussed earlier with the following individual methods:

- *Industry-leading training materials*—The company provides visually enhanced workbooks, pocket-sized job aides, and visceral music-oriented videos.

- *ROCK U*—Hard Rock's online university offers e-learning courses focusing on specific management functions and processes.
- *Rockipedia*—Hard Rock's all-inclusive, online search tool details every cafe process [Hard Rock Cafe, 2013].

Selecting the Training Delivery Method

In order to determine which method is right for your company, you need to decide what type of outcome you want. Based upon that, you must decide which method(s) better support that desired outcome, and at what cost. Once you have that information, you can develop a training plan to support your training needs.

Consultants

While many people are wary of consultants, we know that they can contribute to change (sometimes radical) in a business. These days, most of the successful consultants try to get their clients heavily involved in the process in order to ensure success. If employees don't feel that they were involved and listened to, they will tend not to follow the consultant's recommendations. Many consultants will try to get consensus by having validation of their findings with executives, followed by workshops with key employees, both to confirm their findings and to build support for implementation.

If a consultant expects to be around for the long haul, he or she will still need to do some training and facilitation of events as well. Of course, it is important to have management support (i.e., a steering committee) and the assignment of Lean champions, a kaizen agenda, and so on.

A consultant can help to solidify the opportunities, put together a rough plan, and identify the potential payoff. To really make things happen and stick, you need to have your people trained and involved in the change process.

Training—Key Management Team (Seminars, Certifications, and the Like)

There will need to be some level of experienced leadership if Lean is to be successful. As Lean is relatively new to retail and wholesale, there aren't that many people who have experienced it in this area. So, for the most part, a company will need to develop its leaders, or agents of change, on its own (or use employees with Lean manufacturing experience to help in the process). The best path in this case is to select and develop key employees in your operations and supply chain functions via external training (or possibly use external trainers brought in-house) for very specific types of training that can, in some cases, lead to some kind of certification or educational credit. There are many programs available using traditional methods (attending outside training programs and seminars) as well as via the Web.

As an introduction, there are many off-site (and on-site) seminars for various types of Lean training in manufacturing, distribution, and services. These are typically good starting points to get everyone to a basic understanding of Lean concepts, and in some cases they can be offered to the entire company (e.g., Introduction to Lean).

To truly move things forward, it can help to have key employees take courses that may involve a certification process having different levels of accomplishment, such as a Green Belt or Black Belt in Lean, Six Sigma, or even Lean Six Sigma. One of the best-known programs comes from www.villanovau.com, which is a more traditional university offering various levels of Six Sigma and Lean Six Sigma certification programs. Just do a search online for "Lean training" or "Lean certification" and you'll find an almost endless list of possibilities.

Once these key employees have been trained, they can work with the executive team to establish training and kaizen event schedules for the organization.

Training—General Workforce

Just look up "workforce training grants" for your state on a Web browser, and you'll find that most states have some kind of grant program for free training (typically, you'll still need to pay your workers while they're in training). The purpose of these types of grants is to improve the skills and knowledge of the workforce and at the same time help goods and services companies to grow and prosper so that they hire more people in your state. It's really one of the best things a state can do to help businesses survive in today's competitive environment.

Most of these programs use trainers and consultants from the private sector to deliver the "free" training, but in many cases they also allow for OJT delivered by qualified company employees. The grant application process and deciding which training to include can sometimes be more challenging than the actual training itself. To help with this, in many cases, there are many nonprofit and for-profit companies that will assist with the grant application and management process (beware: some, but not all, of them will charge for this service).

In my experience, doing Lean training via a grant can have various outcomes. Many business owners look at this as free training and don't really have a plan to implement what is taught. This is a big mistake; it usually results in some temporary improvements, but no long-term results. The management must, of course, commit to a Lean transformation, as mentioned earlier in the book, but it must also develop a plan for Lean's successful implementation and assign responsibility for its results.

So while the trainer is both a trainer and a facilitator, the company is responsible for making sure that the proper people are involved and responsible, with clear goals and objectives for outcomes.

In the area of Lean retail and wholesale distribution, I've noticed that there is such an emphasis on productivity (cases per hour, for example) that there is little time available for training operators. It is critical that time be allocated in the annual budgeting process for training of this type; otherwise, you will end up training small groups of people in very short bursts, which isn't very effective over the long term. Again, from my experience doing Lean in the warehouse/distribution environment, it seems that often very little time has been put into the annual budget for training purposes, especially training of this nature.

Obviously, you can bring in outside trainers on a self-pay basis, instead of using some kind of workforce training grant, as they're not hard to find (type "Lean trainers" into any search engine). The advantage of this method is that you can have more control over the content and you don't have to worry about minimum class size, number of training hours, and other restrictions that may be required by your state.

Training—Tools and Tips

Having attended a fair number of courses and seminars on presentation skills over the years has been a great help to me when teaching and training, both in a university and in a business setting. One of the important things that you learn is that people learn through a variety of their senses . . . visual, auditory, and touch or feel. Most people learn best through a combination of all three, but depending on a combination of things such as education level, interest, and job requirements, the balance may vary.

Games

If you're training an executive group, you may be OK with mostly visual presentations (e.g., slide shows and some video examples). If you're working with frontline operators, who are "hands-on" types of people, then you'll need to either have some training simulation games or actually go out into the workplace and implement change.

When you're doing an introductory-type course, it may not be possible for you to actually go out on the floor, so you're best off using a Lean simulation game. If time is limited, then something like the "paper airplane" simulation may be good enough as well as being cost-effective and a great team-building exercise (e.g., the flow simulation found at www.enna.com). In many cases, it may be best (especially with operations people) to use a more specific type of training simulation tool, such as the one mentioned earlier that has been specifically developed for Lean supply chain and logistics management (http://www.enna.com/lean_supplychain/).

The train-do method is very effective, as it uses a combination of classroom and "on-the-floor" training, allowing for a blended mix of visual, auditory, and touch types of learning.

Handouts and Forms

It is also helpful to limit handouts given to the audience before the presentation, as participants have a tendency to look down and read the handouts while you're talking, which can distract them from the learning process. You can always make the handouts available to them afterward, as the handouts may include another learning tool, forms, that may be needed for the hands-on part of the training.

There are many types of forms in Lean, including those used for:

- *Planning*—Team charter and value stream map implementation plan
- *Gathering data*—VSM data collection form, checklists, and activity and process charts
- *Assessing the current state of an area*—Lean opportunity assessment and 5S audits

Language Barrier

A challenge that is becoming more common these days is a language barrier. In manufacturing, wholesale, and retail in the United States, we find a variety of languages spoken other than English, primarily Spanish. Unless the trainer is bilingual, it's best to find someone in the audience who is bilingual and can act as a translator of sorts. It is also helpful to be able to have your handouts available in Spanish. If that's not the case, there are some Lean videos (especially on the subjects of Introduction to Lean and 5S) and handbooks that are available in Spanish (e.g., *The Lean Manufacturing Handbook* by Kenneth W. Dailey and the *Lean Pocket Guide* by Luz Blanco Palma, both found at www.amazon.com).

In many cases, companies have a fairly large part-time workforce, which, again, may include many foreign-language-speaking people (in the United States, Spanish is most common, of course), and detailed training may not be feasible or practical in this case. The use of the previously mentioned videos as a kind of orientation for new hires and temporary employees is helpful. The other thing you can do in this type of situation is to have plenty of standardized work, such as laminated job instructions (in English and Spanish), and a very visual workplace right down to marked floor assignments to make sure that the work is easily understood and followed.

All of these tools and methods must be explored in order to make sure that everyone is properly trained and involved in the Lean transformation process. It may vary in each company based upon the company's specific goals and objectives.

Measuring Success

From a pure training perspective, it is always useful to survey the participants on their perceptions of the course and the trainer. I have also had clients ask that participants be tested in order to make sure that there was a clear understanding of the concepts and applications, which isn't a bad idea. One can use standard quizzes or, for more fun, make a game of it. There is a Lean Jeopardy game available on www.amazon.com and www.theleanstore.com

that can be modified to meet your particular needs (i.e., manufacturing, retail, or wholesale), as it is in slide show format. Playing a game like this is fun and is a great way to reinforce participants' learning and retention.

The benefit of the train-do method is that no matter which method or methods you use for the classroom training, you still end up applying the concepts out on the shop floor to reinforce the classroom learning and get real results with greater enthusiasm through team-building exercises such as simulations and games.

Employee training (and how you go about it) is one of the keys to success in your Lean journey, so it shouldn't be taken lightly and should always be part of an ongoing process, not just a onetime occurrence.

In our next chapter, we will examine how you know whether or not you're on the right track once you've made the commitment and started down the path to becoming a Lean organization.

CHAPTER 15

Defining and Measuring Success: Measurements and Current Statistics

As the saying goes, "You can't improve what you don't measure." This definitely applies in the case of Lean thinking in retail and wholesale.

Any retail performance measurement tool should include data analytics and measurements to help you understand, measure, and track the in-store (and distribution) performance of your own and your competitors' products. You should be able to "slice and dice" the information over time, by customer, and by market.

This information (preferably provided by software) adds value to your business by:

- Tracking key business drivers to identify opportunities and measure progress against objectives through the collection, extraction, cleansing, and presentation of key metrics.
- Removing the non-value-added work of manipulating raw data from your team; this delivers significant time savings and frees account managers to focus on driving sales and profits in their accounts.
- Providing executional and planning insights in a commercial format to leverage with customers.

Interestingly, a report from Ventana Research [cited in *Trends and Outliers*, 2012] suggests that only 34 percent of retail companies are satisfied with the processes they use to create analytics. Possibly contributing to this is the fact that 71 percent of retailers are still using spreadsheets as their primary data analysis tools, the research notes.

Many retailers are paralyzed by the huge amounts of data that are now available to them (remember, data need to be converted to usable information). One example, from a survey, showed that 32 percent of retailers don't know how much data their companies store, and more than 75 percent don't know how much of their data are unstructured and can't be analyzed in a database (e.g., call center notes, online forum comments, and the like).

The survey also found that lack of data sharing is the largest obstacle for more than 51 percent of retailers. Other obstacles include not using data effectively to personalize marketing communications (45 percent) and not being able to integrate data at the customer level (42 percent) [*Trends and Outliers*, 2013].

Key Performance Indicators

Key performance indicators (KPIs) help an organization define and measure progress toward its organizational goals. These goals and metrics can include Lean indicators that are often dispersed in various financial, marketing, and operational areas.

Some of the top KPIs and metrics in retail are [www.klipfolio.com, 2013]:

- *Conversion rate*—Among the most important KPIs for any organization to monitor and measure is the conversion rate of shoppers to actual buyers.
  - *Calculation*—Number of visitors/number of customers
- *Sales per square foot*—This measures your ability to sell a certain amount of product per unit of space. It can also be used on a Web site or at a store to determine how best to place items so that they sell at the best rate possible.

- *Calculation*—Total net sales/total floor area

- *Average purchase value*—This retail metric tells you the average value of each purchase made by one of your customers and varies depending on type of products or services you are selling. This metric is commonly coupled with units per transaction.
  - *Calculation*—Total sales revenue/number of customers or transactions
- *Cost of goods sold (COGS)*—COGS is critical in determining markup percentages for products and in understanding your profit margin. Even though sales may be good, your operating, production, storage, and labor costs may reduce your profitability.
  - *Calculation*—Beginning inventory + purchases – ending inventory
- *Online purchases*—Online purchases versus in-store purchases. In today's world, most on-site retailers have an online store as well, so this metric is used to compare the value and performance of the two purchase sites.
  - *Calculation*—Online purchases compared to in-store purchases
- *Incremental sales*—This measures the amount of incremental sales received as a direct result of marketing activities (e.g., promotions, social networking, direct mail, and in-store promotions). It is important to measure the impact of those campaigns to decide whether they are performing adequately.
  - *Calculation*—Revenue generated by marketing initiatives – baseline sales
- *Customer experience*—This is a broad metric used to determine what areas of your operation satisfy your customers, resulting in their coming back again and again, and what aspects can be improved. This "voice of the customer" is increasingly being used to rate the quality of Web sites, too.
  - *Calculation*—Rate the responses of customers to a given question
- *Customer retention*—This measures your ability to keep customers as well as how viable your business is over the long term. This type of measurement has become easier with the growth of customer loyalty programs such as points cards. You also need to consider attrition and conversion rate data.
  - *Calculation*—25/100 = 25% customer retention
- *Percentage of out-of-stock items*—This measures your ability to keep items in stock and anticipate customer demand. Retailers, which in a way are to a great degree supply chain organizations (plus, of course, carrying out the functions of merchandising, pricing, store location, and layout), need to pay particular attention to their supply chain metrics and KPIs.
  - *Calculation*—# of items out of stock/# of items in stock = % of out-of-stock items
- *Inventory-to-sales ratio*—This compares the amount of on-hand inventory to the amount of sales over a given period and is about striking a balance between sales and inventory.
  - *Calculation*—Average inventory value/average sales revenue

Dashboards to Display and Control Metrics

A very common way to measure, analyze, and manage supply chain performance is with the use of a *dashboard*. The dashboard can be as simple as manually collected data put into a spreadsheet with some graphs, or be a more automated, visually pleasing dashboard generated by an enterprise resource planning (ERP) system. A dashboard helps in decision making by visually displaying leading and lagging indicators in real time (or close to it) from a retail and wholesale process perspective.

Indicators

KPIs are typically displayed in performance dashboards. They usually fall into one of three categories:

1. *Leading indicators*—These have a significant impact on future performance by measuring either current-state activities or future activities.
2. *Lagging indicators*—These are measures of past performance, such as various financial measurements or, in the case of retail, measurements in areas such as cost, quality, and customer service.
3. *Diagnostic*—These are areas that may not fit under leading or lagging indicators, but that indicate the general health of an organization.

Review Scorecard during S&OP

It is important to review these metrics in as close to real time as possible and as part of the sales and operations planning (S&OP) process that we described in Chap. 8.

The benefits of this type of KPI metrics framework are better aligned with corporate strategies and objectives, better collaboration internally and externally with customers and suppliers, an increase in productivity, and greater commitment to and ownership of metrics and targets.

Relevant Lean Retail and Wholesale Supply Chain and Logistics Metrics

The SCOR model (available at www.supply-chain.org) that we discussed earlier in the book can also be integrated with your retail and wholesale supply chain metrics as they relate to Lean. SCOR has come up with five performance attributes, all of which can be related to various forms of waste. They are delivery performance, responsiveness, flexibility, cost, and asset management.

Delivery Reliability

Under the category of delivery reliability, we can look for waste in terms of shipping the correct product to the correct place and customer (or store, which we will use interchangeably in this discussion) at the correct time. This also includes looking at whether or not we have shipped the product in perfect condition and packaging, in the correct quantity, and with the correct documentation. The resultant metrics to look at would include:

- *Delivery performance*—Did the product both ship and deliver to the customers when they originally wanted it? Some companies adjust the delivery date based upon availability, change the date in their system, and measure performance based upon the new delivery/promised date. This results in an inaccurate view of delivery performance.
- *Order fill rate*—It is important to know whether an entire customer order was shipped. This metric typically shows a lower percentage performance than line item fill rate, which should also be measured.
- *Accurate order fulfillment (at various levels of detail)*—This is a quality measurement that looks at shipping errors, such as the wrong order or item(s) being shipped to the customer or store (or the wrong quantity of requested items).

Perfect Order Measure

The culmination of this is the *perfect order measure*, which calculates the error-free rate of each stage of a purchase order. This measure should capture every step in the life of an order. It measures the errors per order line. So, for example, suppose we had the following measurements:

- Order entry accuracy: 99 percent correct
- Warehouse pick accuracy: 99 percent
- Delivered on time: 95 percent
- Shipped without damage: 98 percent
- Invoiced correctly: 99 percent

Our perfect order measure would be only *90.3 percent* (99% * 99% * 95% * 98% * 99%).

This can be a challenging goal to meet when it is set at a high level, but it is a valuable form of measurement that points out the interrelationships among different aspects of your supply chain and gives a good idea of how Lean your total supply chain really is.

Responsiveness

Responsiveness measurements relate to how quickly your supply chain and logistics function can deliver products to the customer. They can include measurements such as order fulfillment lead time, transit times, on-time delivery, and even overall cycle or dock-to-dock time (the total time that key material sits in a facility, which is a good measure of how Lean your organization is).

Flexibility

This is a measure of your supply chain's agility and response time when there are changes in the supply chain. As we know, there can be many unanticipated changes as a result of economic, environmental, political, and other issues that make this something that can be used for a competitive edge.

Cost

It is, of course, important that you manage your supply chain and logistics costs, as that is a sign of potential waste. These measures would include COGS, total supply chain and logistics cost (in dollars and as a percentage of revenue), transportation and distribution costs, warranty/return costs, and a host of other individual costs.

Asset Management

These metrics look at how effectively a company manages its assets to meet demand. This includes both fixed assets and working capital. Metrics include order-to-cash cycle, inventory turns, and asset turns.

Balanced Scorecard

A balanced scorecard (BSC) is a strategic planning and management system that organizations often use to align their business activities with the vision and strategy of the organization, improve internal and external communications, and monitor organization performance against strategic goals.

Today's updated version of the balanced scorecard transforms an organization's strategic plan from a somewhat passive document into the sometimes daily directives for the organization. It provides a framework that shows performance measurements, helps planners identify what should be performed, and measures the organization's success.

A balanced scorecard looks from four different views of the business:

1. *Financial*—To succeed financially, how should we appear to our shareholders?
2. *Customers*—To achieve our vision, how should we appear to our customers?
3. *Internal business processes*—To satisfy our shareholders and customers, at what business processes must we excel?
4. *Learning and growth*—To achieve our vision, how will we sustain our ability to change and improve?

Objectives, measures, targets, and initiatives are developed for each of the identified perspectives to ensure success.

Retail Balanced Scorecard Examples

Tesco BSC Example

Tesco, the third-largest grocery and general retailer in the world, uses something they call their "steering wheel," a balanced scorecard approach, which measures their success and is an important tool to gauge performance and to steer their business in the right direction.

It was rolled out to all departments. Stores have a visual display of the steering wheel, and key measures behind it would be updated every week for staff to work in a common direction.

This also is complimented by a visible and effective communication plan for the steering wheel, which includes the display of the wheel, the summary and actions, and the change projects initiated to keep the measures performing as desired. It is a visual way of measuring their performance, of seeing how they are doing and where they need to focus.

This is reviewed every month and is linked to the corporate steering wheel. For example, they cannot think about reducing procurement costs, while reducing the quality of products they offer.

The scorecard has five segments (Customer, Community, Operations, People and Finance), and each segment can have multiple objectives. They have a number of measures that support these objectives which are reviewed every year to ensure they are aligned to the core strategy.

They are communicated and updated for everyone to be engaged. Ownership is top down but the objectives and measures are part of an individual's objectives. Finally, weightings must be given to critical measures to ensure they get more focus when underperforming [Thomas, 2011].

Wildcat Inc. BSC Example

Wildcat, Inc., [the fictitious name for] a national merchandising company, sells a specific type of tangible property as its primary business and operates food and beverage services with a limited menu. The company is one of a small number of merchandising companies that control a large portion of this product's retail market. . . .

Prior to its BSC implementation, Wildcat added performance measures as it needed to solve a particular problem or issue. This led to a performance measurement system that was neither concise nor well structured. Throughout these changes, top management focused on two critical-success factors: actual sales relative to budgeted sales (sales-to-plan variance) and inventory management. . . .

The system represents an attempt to control and compare the results of management at the store, district, and regional levels.

Wildcat's BSC currently consists of measures that evaluate store performance. The measures for relevant stores are then aggregated to assess the performance of district and regional managers. Wildcat's choice of measures reflects upper management's desire to promote its annual goals even though Wildcat's culture incents managers to primarily focus on one measure: sales-to-plan. . . . The measures included in the performance system are:

1. Sales-to-plan (a comparison of actual sales to budgeted sales)
2. A customer service evaluation
3. Payroll percentage
4. Shrinkage
5. A company-specific measure of store efficiency in areas such as supply chain management, communications, and service
6. In-store audits [Biggart et al., 2010]

At this point in the book, you should have everything you need to know in order to start (or continue) the Lean journey in your retail or wholesale organization by using a holistic approach that offers a long-term, permanent solution. So, on a final note, let's look at what's lies ahead so that we can anticipate things that could get in the way of our success.

CHAPTER 16

The Road Ahead: Thoughts and Suggestions on the Future of Lean in Retail and Wholesale

Throughout this book, I've tried to give the reader knowledge concerning concepts, tools, and examples of Lean for retail and wholesale. Sometimes the biggest challenges can be on the "soft" side of business in terms of an organization's culture and commitment. In other cases, it may be a question of how or where to get started. No matter where your organization is, not only do you need to start thinking Lean if you are to become more competitive, but you also need to anticipate trends if you are to stay competitive.

People, Process, and Technology Views

As I mentioned in Chap. 1, when you are thinking of implementing Lean (or any process improvement program, for that matter) in your organization, I think it is effective to think of both the current *and* the future state of an organization in terms of people, process, and technology, which is a good way to start this discussion.

People

The success or failure of a business depends on its employees. If they aren't part of an improvement program, it is doomed to failure. Whether you're a manufacturer,

a wholesaler, or a retailer, there has to be both top-down support and bottom-up involvement.

In addition to the "What's in it for me?" question, employees want to feel that they are listened to and are part of the solution, not just viewed by management as being part of the problem.

It's useful to think of a Lean program in terms of Maslow's hierarchy of needs, which states that our actions are motivated in order to achieve certain needs. If we are successful in meeting many of these needs, our employees will tend to be more productive and willing to participate, both in general and in terms of a participatory program like Lean.

Specifically, Maslow's hierarchy of needs, when viewed in "pyramid" form (Fig. 16.1), points out that people have basic physical requirements, including the need for food, water, sleep, and warmth. Once these lower-level needs have been met, they can move on to the next level of needs, which is for safety and security. As people progress up the pyramid, their needs become increasingly psychological and social. Thus, the need for love, friendship, and intimacy becomes important. Further up the pyramid, the need for personal esteem and feelings of accomplishment takes priority.

The more we can help our employees to progress up this pyramid of needs, the more motivated they will be, and the more successful the organization as a whole (and the Lean program) will be.

In terms of the success of a Lean program, as we've stated throughout the text, we need to support (financially, emotionally, and educationally) and empower our employees if we are to fully motivate them.

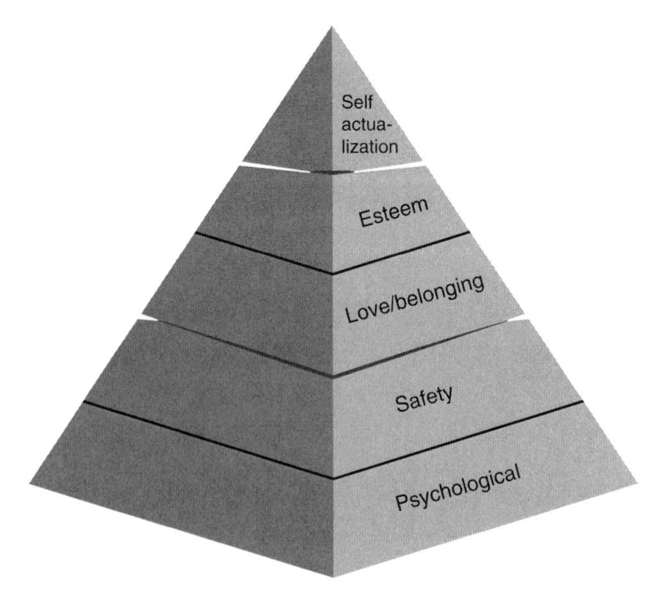

Figure 16.1 Maslow's hierarchy of needs.

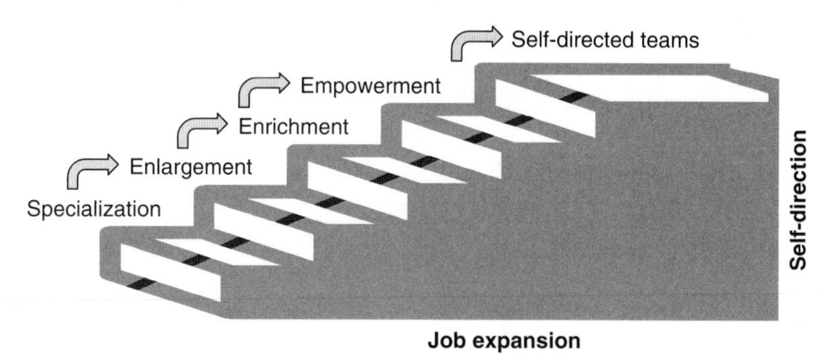

Figure 16.2 The job design continuum.

Something that is known as the *job design continuum* (Fig. 16.2) illustrates how we can progress in this task of self-actualization in the workplace (i.e., to the top of Maslow's hierarchy of needs).

As we move up the job continuum by enlarging and enriching jobs, we reach a point at which we can empower employees, which is sometimes a very difficult task for some managers to deal with, as they may have to give up some control. Empowerment is the process of enabling an employee to think, behave, act, react, and control his or her work in more independent ways.

To be truly successful in Lean (and any type of program or project) we need to establish teams (in some cases self-directed, but not necessarily), as was pointed out in Chap. 12. This ultimately leads to benefits such as an improved quality of work life for employees, improved job satisfaction, increased motivation, improved productivity and quality, and reduced turnover and absenteeism.

Process

The number of stock keeping units (SKUs) offered at retail has exploded over the past 40 years or so. The average supermarket today carries anywhere from 15,000 to

60,000 items compared with 5,900 in 1960 (and 7,800 in 1970) [McTaggart, 2012; Reinvestment Fund, 2011].

As a result, it has become harder and harder for wholesalers and retailers to manage their operations. That is perhaps why Lean in retail and to some degree in wholesale started with programs like Efficient Consumer Response (ECR) and Collaborative Planning, Forecasting and Replenishment (CPFR) to help firms better manage their inventory and the inventory in the supply chain.

Additionally, in these tough economic times, retailers are looking for any edge they can find to improve their thin margins and increase productivity and sales. That is why throughout this text I have advocated the need to look beyond the supply chain (which is a great place to start) and include all of the processes found in retail and wholesale, not just in operations, but also in sales and marketing, merchandising, and even finance.

Technology

Technology can enable a good process. The past 20 years have seen exponential gains in technology, including the growth of the Internet (for e-commerce

and m-commerce, collaboration, and visibility), enterprise resource planning (ERP), supply chain planning and execution systems, and warehouse and store technologies (e.g., radio-frequency identification (RFID) and self-checkout).

It is important to stay up to date in terms of technological advances, as they are coming at us at such a rapid pace. To do so, let's take a look at some of the current trends that we see.

Trends

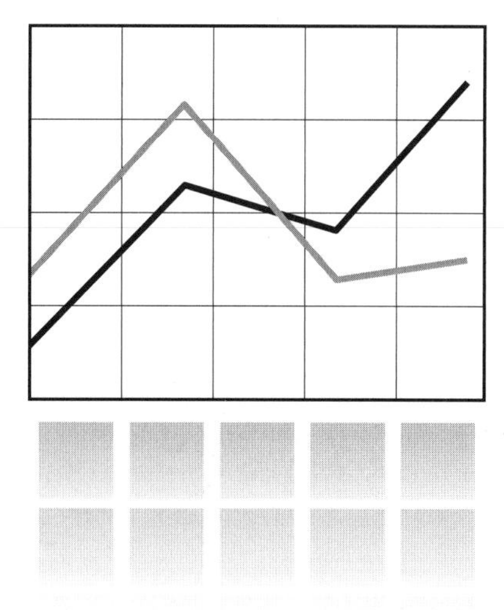

Retail

There are a number of trends in retail (some of which we've pointed out earlier in this book) that are critical to future efficiency and profitability, as pointed out by *What's Next* [*What's Next*, 2013]:

- *Self-serve*—Everyone saves money, and customers think they're in charge. Examples include self-check-in kiosks in airports, self-scanning checkout machines in supermarkets, and DIY (do it yourself) checkout services in hotels. Intelligent vending machines are coming in the near future, and there's even a car dealership in Japan that uses robots as salesmen!
- *Blurring of sectors*—We now see bookshops that sell coffee and food, coffee shops that sell music, and supermarkets with banks that offer loans. This is an opportunity for sales growth, but managing it efficiently is a challenge.
- *High-speed retail*—People have increasingly hectic lives, which is driving trends like e- and m-commerce, drive-through and take-out dining, and mobile banking. This faster-paced shopping is also hurting malls, as it takes longer to shop there. Because of people's ever-shortening attention span, we are also seeing more "pop-up" retail stores and "limited time only" products and offers.
- *RFID*—As we've previously discussed, RFIDs contain microchips with an antenna that retailers can use remotely to manage their inventory and prevent theft. In the future, RFID readers may be used to scan bags as they leave a

store and automatically take the money from a suitably enabled mobile phone linked to your bank or credit card.

- *Mass customization*—This might mean tailor-made products, limited-run products, or mass customization (products created for specific niches or groups, often with the help of the customers themselves). Mass customization is a trend created by "commodification" (i.e., the transformation of goods or services into commodities), which has, in turn, been created by globalization.

Other trends that are and will be affecting retail efficiency and sales for the next 5 to 10 years, as pointed out in *Chain Store Age* [Welty, 2013] include:

- *Social networks*—Using analytics tools, retailers can mine these data to gain insights into what consumers want to buy and use these insights to create more effective merchandising plans. Also, retailers can use social networks to build relationships and attempt to influence what and how consumers buy.
- *Showrooming*—This occurs when shoppers go to brick-and-mortar stores to examine merchandise, but later buy it online (we've all done it at some point in our lives!). While many brick-and-mortar retailers don't like this, they need to look at it as an opportunity to engage in-store shoppers to increase revenue and loyalty, both online and in physical channels. To do this, retailers will have to better train and equip store employees and consider new service-based store layouts with "curated" assortments, such as organizing items into branded sections and grouping products according to themes (often based upon consumers following the new "curators of style and taste," such as Martha Stewart). They will also have to allow stores to have increased autonomy in order to leverage local events, negotiate prices, and cater to local demographics.
- *Store-size rationalization*—Retailers will need to review their real estate strategy and understand the new role of physical channels in the all-channel commerce. They may need to adjust store sizes, layouts, assortments, fixture arrangements, and "entertainment" factors to deliver the total brand experience.
- *Near-field communication*—Retailers must increase the use of communication technologies in their stores to gain a better understanding of how shoppers move through the store to make purchasing decisions and where their interests may lie, and offer them contextual, real-time options using mobile apps and interactive store displays.
- *Growth in order fulfillment from stores*—Customers now have the ability to order online and then pick up the merchandise from a local store (e.g., Walmart's "Site to Store" offering). Traditional retailers that use this form of delivery need to reexamine their policies on staffing, compensation, task assignment, inventory planning, and sales reporting. They also need to determine whether this cannibalizes inventory at the expense of foot traffic and how to measure and compensate employees for sales made online but fulfilled in the store.

The *Chain Store Age* article also points out that to leverage these trends, retailers need to break down the demand silos, create a single, aggregated demand forecast for all channels, and have supply chains that are responsive and flexible, (i.e., sales and operations planning (S&OP)).

Furthermore, as retailers evolve their supply chain operations from single-channel to all-channel order demand, processes become more complex, which can escalate costs. As a result, retailers will need to balance the cost and service variables in their inventory management operations. To be successful, they need a real-time view of orders, inventory, and expected deliveries from suppliers.

These trends all point out the need to be more efficient, which these days is "top of mind" to retailers, according to a recent survey [Vanson Bourne and NetSuite, 2012], in which the researchers found that "increasing profitability is, unsurprisingly, a top priority. Although most retailers have already gone through belt tightening processes as the recession hit, more than half (52%) still see further opportunity to cut costs, with streamlining of general business processes (43%) and improved processes and systems within their supply chains (43%) also identified as opportunities to make more profit. More than one in five (22%) identify cutting costs as their biggest opportunity."

As we can see, Lean can make both a short- and a long-term contribution to retailers in this regard, as described throughout the book.

Wholesale Distribution

There are a number of trends in wholesale distribution that can have an impact on current and future efficiencies in the business, according to a recent report by *Modern Distribution Management* magazine [Konzek and Stulton-Holtmeier, 2013]:

- *E-commerce is gaining critical mass*—Business-to-consumer practices are being transferred to the business-to-business world in areas such as online catalogs and online ordering, and small-to medium-sized distributors can no longer afford to ignore this trend.
- *Mobile is becoming a "game-changer"*—Distributors not only are using m-commerce as a platform for purchasing and access to their customers, but are also using mobile devices to allow their salespeople to be more productive.
- *Analytics is taking center stage*—Distributors are now in a position to use the data in their systems to quickly look for trends and opportunities for growth. The real benefit of this will be to use these data to improve processes and gain a competitive advantage.
- *Distributors are facing a competitive shift*—E-commerce, m-commerce, and other technological advances have allowed distributors to compete more effectively and through more channels, but they've also allowed for the entry of new competitors.
- *Vending is continuing to grow explosively*—There is a greater use of vending by a broader range of distributors, especially in vendor-managed inventory (VMI)-type applications (e.g., vending machines with tools connected through the Internet that are placed in the customer's facility). While many distributors are

against this type of business model, many customers want easy and accurate control over products that can be supported by vending solutions.

- *More distributors are going global*—Distributors in many sectors have announced acquisitions or joint ventures globally. While this can result in increased sales, it can create a more complex and harder to manage supply chain.

The aforementioned Vanson Bourne study [Vanson Bourne and NetSuite, 2012] shows that wholesale distribution companies' priorities for improving profitability also include cutting costs and better use of e-commerce capabilities (unlike among retailers, having an e-commerce site is not a given in the wholesale distribution industry)—both areas where Lean can help.

The study mentions that the cost of customer support and service is seen as the biggest challenge for wholesalers in terms of increasing profitability, in addition to the cost of marketing and customer acquisition. Some other areas that are critical challenges for wholesalers are improved demand forecasting and communications with suppliers (30 percent were still communicating with suppliers by manual or telephone ordering).

Wrap-Up

The time to start thinking about the future is *now*. History tells us that many businesses wait until it's too late and eventually fail as a result. In many cases, this could have been avoided if they had had a Lean culture in place to make sure that they had left no stone unturned by listening to the voices not only of their customers, but also of their employees.

If we look at a "best in class" grocery retailer like Wegmans that uses continuous improvement tools in practically every aspect of its business and compare it to Pathmark (both of them located on the East Coast of the United States), we can see a stark difference.

A 2012 *Consumer Reports* survey [*Consumer Reports*, 2012] of its subscribers rated Wegmans as one of the top grocery stores in America. Survey respondents felt that it offered high-quality meat and produce, a clean shopping environment, and very good or exceptional prices.

Pathmark's ratings in the survey were much lower. Pathmark, now owned by A&P, received "highly dissatisfied" ratings on service, mediocre ratings on food quality and price, and poor ratings on cleanliness. Among the shoppers surveyed, 75 percent had one or more complaints, and 31 percent had experienced three or more problems with Pathmark.

The types of issues that the survey found at many of the lower-rated stores included not enough open checkouts, congested or cluttered aisles, and advertised specials that were out of stock. The survey also found inept bagging, missing prices, and scanner overcharges.

I'm sure at this point the reader recognizes that all of these types of issues would be less prevalent in a store that had implemented a continuous improvement program.

Conclusion

While implementing a Lean continuous improvement program in an organization is neither fast nor easy, the reader should now have a good idea of its value and an understanding of what it takes to be successful.

It's now in your hands to decide whether you want your company to use Lean to become a "road grader" and flatten the competition or wait until it's too late and become "road kill" instead. When put in those terms, it's not such a tough decision after all, is it?

APPENDIX A

Case Studies

Retail Strategy: Getting Lean in the Retail Sector—Bob's Stores Makes It Happen*

Introduction

Based in Meriden, Connecticut, Bob's Stores is a chain of more than 30 retail apparel stores located primarily in the northeastern United States. Founded in 1954, the company offers name brand clothing and footwear. Facing strong competition, the company launched an organization-wide improvement initiative to enhance the overall customer experience.

**Source:* Maynard, "Bob's Stores Case Study," http://www.hbmaynard.com/CaseStudies/ArchivedCaseStudies.asp.

In particular, the management team at Bob's wanted to build a lean organization with a culture focused on customer value and waste reduction. They also wanted to create a culture of continuous improvement in which store associates would embrace change, have pride in their workplace, and raise what already was a high level of service.

The improvement initiative was conceived with three key objectives in mind:

- *Appropriate staffing*—Bob's wanted to improve its ability to forecast staffing requirements by using engineered standards, rather than sales volume.
- *Checkout efficiency*—The company wanted to identify and eliminate unnecessary activities that hindered smooth transactions at checkout lines, associates to spend less time on checkout tasks and more time assisting customers.
- *Enhanced service*—Bob's wanted to improve the customer experience by making it faster and easier for customers to locate items on the sales floor.

To meet these objectives, Bob's engaged H.B. Maynard and Company, Inc. to help assess the company's current state and provide expert assistance and training.

Using the 5-S Process

Maynard's assessment revealed that several improvements could be made through better work methods and engineered standards. More importantly however, the assessment showed that such improvements could not be realized unless employees embraced a culture of continuous improvement.

To accomplish this, Maynard recommended a 5-S initiative, which would not only improve orderliness and eliminate unnecessary activities, but would also establish an environment prepared for further improvement.

5-S places an emphasis on and prepares an organization for lean operations. This requires a culture that is willing to change and review existing processes and methods, with a focus on the reduction of waste. 5-S helps to create an environment that is clean, orderly, and safe. It also begins to open the company culture toward change and instills new discipline.

This new culture must focus on the importance of customer response time and value as defined by the customer. Once established, this new culture becomes the foundation for further workplace improvements.

Launching the 5-S Initiative

Beginning with a pilot project at one store, Maynard worked with Bob's to develop a team of associates skilled in the disciplines of 5-S, best methods and engineered standards. A cross-section of the sales floor, backroom operations, and common areas were chosen to demonstrate that 5-S concepts are important to all areas of the store.

To kick off the project, Bob's established a comprehensive roll-out plan involving several teams consisting of store and home office associates. The teams were schooled in 5-S concepts and techniques by taking a half-day training course at Bob's corporate office. Because all teams took part in the training, there was a heightened level of interaction among senior managers and sales floor associates, which reinforced management's commitment to the project.

Let the 5-S Begin

With training complete, Bob's management and sales floor associates began working with Maynard consultants to tackle the first "S"—sort and remove. The step involves the removal of all unnecessary items, leaving only those that are necessary to perform the work.

At the cash register, for example, many unnecessary items—broken hangers, old store flyers, and broken bag stands—were tagged and removed to a holding area, repaired or simply discarded. Items were cleaned out of the holding area weekly, and decisions were made about where to place the items.

In the store's backroom, the sort and remove process yielded even greater benefits. Through the process, store associates cleared off 42 8-foot shelves, providing Bob's with 1,260 cubic feet of shelf space to either store merchandise or reconfigure for a more efficient backroom layout.

"5-S helped us to improve space utilization significantly in the store," said Bob's Stores' Senior Vice President of Store Operations Scott Hampson. "This allows us to process and store additional inventory more effectively. This is critical during our peak selling seasons."

Shine and Inspect

The focus of the second "S"—shine and inspect—is to improve the condition and appearance of the work space, to make it look "like new," and to inspect all items for deterioration.

At the cash register stations, associates scrub the work surfaces inside and out to make the counters clean and appealing. Associates also inspect each cash register, including looking for broken keys on the register. While a seemingly small matter, a broken cash register key could affect customer flow at the checkout line.

In addition, the floor of the processing area in the backroom was scrubbed and sealed to maintain a clean, fresh look. This helps to minimize damage to merchandise, as it reduces the chance of clothing becoming soiled if it is dropped during processing.

Set Locations and Visual Cues

This 5-S step involves creating a home location for every item and establishing visual cues so associates can locate items quickly. This process is implemented in several ways throughout the store.

At the store's service desk, associates set alphabetical locations to organize customer holds and returns. Clear bins are used as visual cues, so associates can see what is in the bin as they approach it.

A mapping technique is used to help associates locate products in the backroom. Using a letter and number grid, employees can find products quickly. Vertical space is maximized by using shelving units, reducing the amount of space needed to store merchandise. A color-coded signage system, organized by merchandise category, helps associates locate products more efficiently.

Bob's Stores uses color-coded signage on the sales floor to help "funnel," or direct customers to their desired location. For example, if a man wants to purchase denim jeans, store overhead signage helps him locate the men's department. Once in the department, additional signage funnels him to the jeans section, and to his desired brand of jeans.

Through 5-S, Bob's Stores extended the use of color-coded signage to further benefit customers. Now when the man in our example finds his brand, he uses a color-coded key, located on each of the store's shelving units, to find his size easily. He can quickly find his size because the denim labels are colored to correspond with the key.

Used throughout the store, the color-coding system enhances customer service by helping customers to locate items quickly and easily. It also reduces the amount of time Bob's associates need to spend helping customers locate merchandise in their size, freeing them to help more customers with other needs.

Systematize

This "S" involves instilling habits to ensure the first three steps become incorporated into everyday routines. Techniques applied here include checklists and assignment maps. For example, checklists help cashiers determine what needs to be done daily, weekly, and monthly to keep the front-end area in like-new condition.

Stay the Course

The fifth "S" is stay the course. To encourage store associates to practice 5-S concepts, Bob's offers a recognition and reward program, and communicates 5-S values consistently to associates.

For stay the course, a reward and recognition system was designed to honor associates who exhibit the 5-S discipline in their work area. Bob's senior managers and store associates created Gold, Silver, and Bronze levels which were tied to 5-S achievements. Associates can win free movie passes or other prizes for various levels of 5-S achievement attained through weekly 5-S audits.

In addition, a 5-S newsletter was created and distributed to associates on a regular basis to update them on 5-S practices used throughout the store. A 5-S communication board was also located in the break room, to keep 5-S in front of all store associates.

5-S Results

Because 5-S is an ongoing, cultural process, Bob's expects to see continuous improvement as a result of its initiative. 5-S results thus far have been very favorable. For example, the store:

- Improved efficiency at the front end, speeding the checkout process and allowing associates more time to support customers.
- Effectively used color-coding and signage, allowing both associates and customers to locate items more quickly.

The initiative also significantly improved space utilization. As a result of 5-S tactics, the store:

- Disposed of 20 buckets of scrap metal, freeing up 360 cubic feet of space.
- Disposed of or reorganized fixtures freeing up 1,260 cubic feet.
- Cleared off every top shelf in the backroom to increase available space.
- Removed a trailer from the parking lot that was housing fixtures. This eliminated the need to pay rent on the trailer.

Next Phase

With 5-S implemented in all areas, Bob's will shift its focus on best methods and engineered standards. By first establishing an environment of continuous improvement through 5-S, Bob's will be ready to capitalize on the additional benefits that will result from implementing those processes.

5-S Results for Bob's Stores

Store Area	5-S Results
Front-End Area/Checkout	Faster checkout times; cleaner work area; removal of unnecessary items improves work flow; associates have more time to help customers on sales floor
Sales Floor Associates	Associates can find products more easily; use of color coding helps customers find products
Backroom	Improved space utilization by freeing more than 1,260 cubic feet; additional space used to store merchandise, making the space especially valuable during peak selling seasons

Merchandise Management: West Marine: A CPFR Success Story (by Larry Smith, Senior Vice President Planning and Replenishment, West Marine Inc., March 1, 2006)*

In 1997, my company, West Marine Inc., acquired an East Coast competitor, E&B Marine. The consequences were quickly apparent: sales fell by almost 8 percent, and peak-season out-of-stock levels rose more than 12 percent compared to the prior year. We soon felt the effects on the bottom line: After six years of steady growth, net income dropped from $15 million in 1997 to not much more than $1 million the next year.

Fast-forward six years to 2003, when we purchased our largest competitor, Boat U.S. We successfully integrated their distribution center (DC) in just 30 days and their in-store systems inside 60 days. During the 2003 peak season, we had no supply problems in any of our warehouses or stores, and the acquisition was accretive in the first year.

What changed? Two words: supply chain. Into the late 1990s, West Marine had not fully recognized the value of effective supply chain management. After the tough E&B acquisition, our management team realized we had to make a significant cultural shift. Traditionally run by "boaters first and businessmen second" (as one manager put it), the company now had to be run with discipline. Part of

Source: Larry Smith, "West Marine: A CPFR Success Story," *Supply Chain Management Review*, March 1, 2006.

the transformation involved overhauling our supply chain operations internally and with our suppliers. A crucial element would be the development of a supplier collaboration program based on collaborative planning, forecasting, and replenishment (CPFR) principles.

The company's approach has also led many of our key suppliers to shift their own views of supply chain management beyond the goal of improving retail in-stock rates. For some, this shift has been nothing short of transformative. I'm glad to report that the changes at West Marine and its collaborating suppliers are not one-time events. Together, we have developed a sustainable process of supply chain performance improvement.

This article will chart West Marine's supply chain overhaul, showing how we absorbed CPFR principles and procedures into our operations. Importantly, I'll emphasize how we worked toward and achieved what we might call "full-strength" CPFR—an embrace of everything that the collaborative principle can do to create and sustain an adaptive, high-performance supply chain. I'll also describe the "how" of what we do to make it work—the techniques we used and still use to secure buy-in from suppliers and to ensure that their commitment sticks.

From Anchors to Wetsuits

Founded as a mail-order firm in 1968 by avid boater Randy Repass, West Marine opened its first retail store in Palo Alto, Calif., in 1975. Three years later, the company created a separate sales channel to commercial customers such as boat yards and boat dealers. In 1987, the company published its first retail catalog. Today, with more than 400 stores and annual sales of $690 million, West Marine is the largest boating-supply retail chain in United States. The retailer sells more than 50,000 products through its stores, Web site, catalog, and commercial sales arm—from stainless steel propellers and anchors to life jackets and wetsuits.

The after-effects of the E&B merger alarmed the board. It was clear to founder Randy Repass that West Marine needed deeper management bench strength. In late 1998, Repass and his fellow directors brought in a retail veteran, John Edmondson, as CEO, with Repass as chairman. Edmondson reinforced the senior management team and set in motion sizeable investments in new systems and processes. He hired David Schenk, an 18-year IT executive, as CIO. Pat Murphy came aboard as senior vice president of logistics; I arrived in 1999 to head planning and replenishment. Like my newly hired colleagues, I came from a Fortune-500 company; I'd been leading the supply chain planning team for Kmart Corp.'s $8 billion apparel business.

It was at Kmart that I'd first learned about CPFR, and I could envision its value to my new employer. We'd run some early CPFR pilots at Kmart. But I'd also attended a few conferences of the Voluntary Interindustry Commerce Standards Association (VICS), the body that oversees, upholds, and advances the CPFR standards and owns the CPFR trademark. I'd been to several of VICS' CPFR Committee meetings and presented at a few. Through my VICS activities, I'd met and worked with VICS president and CEO Joe Andraski; with Jim McLaughlin, Procter & Gamble's vice president of process and support services; Fred Bauman, JDA Software's vice president of collaborative solutions; Matt Johnson, then Syncra's chief technology officer; and with other thought leaders who knew plenty about collaborative supply chain practices.

A word about CPFR: It's a business practice that combines the intelligence of trading partners to improve the planning and fulfillment of customer demand. CPFR links best practices in sales and marketing to supply chain planning and execution processes in order to increase availability while reducing inventory, transportation, and logistics costs. The concept crystallized in 1998 around the consensus that it was not enough to coordinate the supply chain within the boundaries of a single organization.

In its first phases, CPFR was interpreted as a linked sequence of nine business processes, backed by industry standards. But in practice, CPFR wasn't as linear as the nine steps implied. So today it is viewed as a wheel that can be entered at any point, as depicted in Exhibit 1 [Fig. A.1]. The implementation does not need to embrace every element to realize value.

The Path to CPFR

Tackling our post-acquisition challenges in late 1999, most of our executive team thought we had four supply chain problem areas: the DCs, transportation, replenishment, and the systems supporting these operations. So logistics chief Pat Murphy set out to establish reliable DC operations and on-time transportation services. CIO David Schenk made sure the existing system infrastructure delivered timely and accurate data processing. I concentrated on our replenishment performance (especially the in-stock levels in our stores), bringing it up to mid-90-percent levels and above—at par with most retailers' results at that time.

As things stabilized during 2000 and 2001, management's perspective on our "supply chain problem" changed. Although our execution improved, suppliers were still having problems filling our orders. We also realized that West Marine was carrying too much inventory to make up for supply shortfalls. Supplier on-time order fills were a dismal 30 percent.

There's a crucial point to make here. Up to 2001, world-class thought leaders in CPFR wouldn't have considered our few initiatives to have been representative of what this technique could achieve. Although I'd begun articulating the benefits of CPFR not long after I joined West Marine, I knew it was a huge step for the management team to take. It would take time—and education. West Marine did approve moderate investments in IT systems to enable CPFR. But it took a further realization—that we actually had a vendor fulfillment problem—to reveal what was constraining our plans for growing the business. It was that realization that caused West Marine's management team to crystallize around the concept that would improve our suppliers' shipping performance.

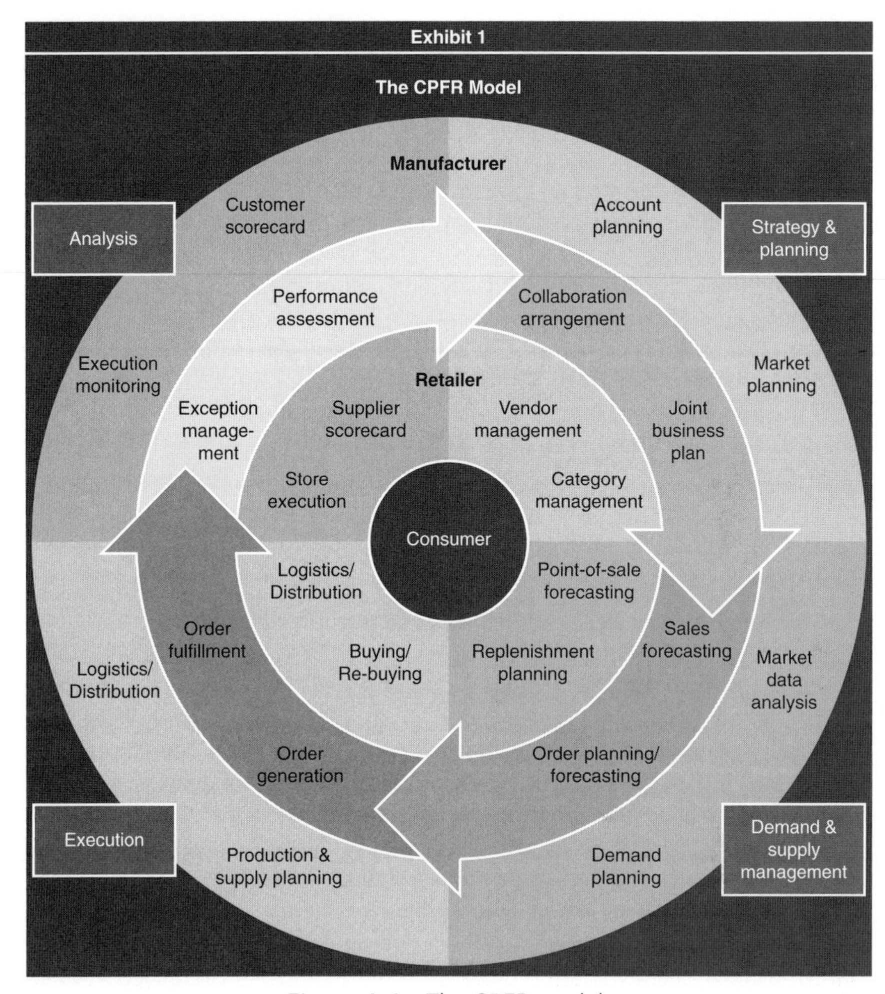

Figure A.1 The CPFR model.

Our first CPFR pilot program kicked off in January 2001. We recognized that for us to become a great supply chain company, we needed to change our core culture as well as execution. What was CPFR's role? The business practice clearly showed how we could improve performance using cross-functional and intercompany processes supported by enabling technology.

Now jump forward to late 2004, by which time we had 200 CPFR relationships, our suppliers' on-time order fills had improved to nearly 80 percent, and we were delivering 96 percent in-stocks during our peak season. Once again, West Marine's management team had to shift its perspective on the company's supply chain problems. We now saw that we had a supply chain planning problem in concert with our vendor partners. We wanted our supply chain to deliver better solutions, programs, products, and promotions. Because of our commitment to CPFR, we had moved from a push supply chain model to a pull model. We had implemented technology programs that established West Marine as a leader in supply chain execution. Yet excellence still eluded us. As our own view of the "problem" evolved, the lines also began blurring between supply chain improvements and business strategy optimization. It became clear that a rudimentary "textbook" approach to CPFR would not be sufficient. In effect, we had to look at, understand, and fully embrace the spirit of collaboration. That might sound flakey, but it's not. It meant addressing cultural issues just as much as business processes.

For West Marine and many of its key suppliers, CPFR is a core business process that provides a path to accelerated performance improvement. Its role is similar to the organizational improvements wrought by corporate and supply chain programs in quality improvement, lean/Six Sigma, the Supply Chain Operations Reference (SCOR) Model, and sales and operations planning. Such programs provide specific process maps and an integrating philosophy that help organizations to become more adaptive and performance-driven. We identified 10 performance-improvement steps that reflect our "full strength" approach to CPFR:

1. Seek long-term, holistic solutions, not quick or myopic fixes.
2. Reconcile conflicting goals and metrics.
3. Pursue inclusive problem-solving; do not depend upon "experts" who don't have accountability for the business.
4. Instill collaborative processes that encourage idea creation, shared problem-solving, and high adoption rates across organizational boundaries.
5. Use a disciplined and iterative set of methodologies such as CPFR, SCOR, or Six Sigma to help teams define issues, root causes, and solutions.
6. Develop a culture of continuous improvement, particularly at the customer-facing associate level, because those employees are most likely to know what's needed.
7. Create clear accountabilities and assign authority with a focus on core business processes rather than on traditional organizational "silos" or loyalties.
8. Commit to technology enablement for execution, communication, exception management, and root-cause analysis.
9. Reduce decision cycle times.
10. Implement rapidly.

While these 10 elements could come from any number of sources, Hau Lee and Jason Amaral's "Supply Chain Management Performance" paper from the Stanford Global Supply Chain Management Forum has been particularly useful. The article refers to the "adaptive organization," which ensures a continuous focus on the right things through responsiveness and balance. The entire organization is performance-driven; it sets goals, addresses root causes, and leaps on competitive opportunities. This is what West Marine is striving to become.

The Right Replenishment System

Technology enablement was an essential precondition for our CPFR successes. To our knowledge, West Marine is the first consumer-goods retailer to implement an aggregate ordering or "multi-echelon" replenishment process. A "multi-echelon" replenishment solution integrates the forecasting and replenishment solutions in a retailer's stores and warehouses. Data such as seasonal forecasts, promotion stock levels, and future assortment changes are automatically calculated so that the retailer gets a reliable projection of its future orders and so that suppliers can continually deliver accurate, on-time orders to the DCs. The resulting order forecast is in the form of machine-readable data that can be entered into a supplier's materials requirement planning (MRP) or enterprise resource planning (ERP) system to represent a key customer's demand. While third-party solutions are available, very few retailers have completed similar installations as of this writing. This is something of an enigma, given how long CPFR has been around and how much it has been discussed. (I'd once predicted that retailers would quickly embrace new replenishment systems. But I think I underestimated the ability of companies to hang onto the status quo.)

Retailers' conventional installations of automatic replenishment and forecasting software run on parallel but unconnected systems at store and warehouse levels. The store system uses actual customer-purchase data to forecast future customer purchases and what the store needs to order to support retail-service levels and in-stock levels. But the warehouse system uses actual warehouse-shipment data to forecast future shipments and what the warehouse needs to order to support their service and in-stock levels. In other words, the warehouse system has no data about store-level overstocks or understocks, and promotional needs must be forecasted in both platforms. Changes in demand—caused by product-mix changes in the store, for example—aren't communicated systemically to the warehouse-level replenishment system. Imagine a situation in which last year's major store promotion will not be repeated this year. Using the disconnected replenishment model, the historical-shipment information might still include last year's promotional-inventory build, and the warehouse-replenishment system would purchase according to last year's demand, thus overstocking the DCs. It happens all the time.

West Marine's multi-echelon replenishment solution resolves the store-warehouse disconnect. Warehouse replenishment immediately responds to all store-level overstocks and understocks. Similarly, all promotions and store-level assortment changes are planned in the store system, and warehouse replenishment immediately responds to them. The solution eliminates duplicate forecasting tasks and creates more accurate supplier-order forecasts (See Exhibit 2 (Fig. A.2)).

In contrast to many CPFR initiatives where ordering is constrained to the forecast, West Marine issues orders in an unconstrained manner each business day, as recommended by the system. That way, we're always purchasing optimally, based on

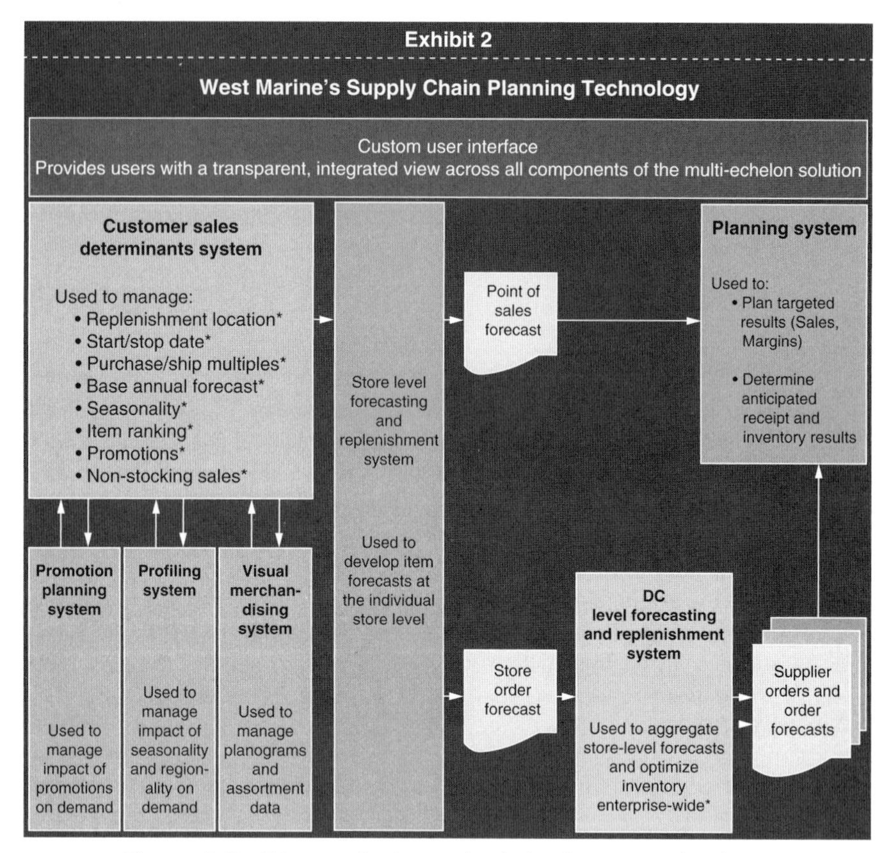

Figure A.2 West marine's supply chain planning technology.

the latest sales and planning information. We set a requirement that all processing needs to be accomplished nightly, recognizing that timeliness is key to accuracy. Fully 97 to 98 percent of the items at West Marine are managed through automatic forecasting and replenishment, and we conduct our entire process, from updating to supplier-order forecasting, every night. Any change in store-level replenishment will automatically be processed in the next day's supplier order and order forecast. Based on the robust forecasting capabilities of our existing JDA/E3 forecasting systems, the supplier-order forecasts we generate daily are by item, warehouse, and week for a year.

West Marine's choice of supporting IT systems sprang from a decision to pursue a particular strategic approach to supply chain collaboration. When we were launching our CPFR effort, many CPFR pilot initiatives involved a limited number of items and suppliers, with both the retailer and the supplier providing

Exhibit 3 CPFR Alternatives			
Alternatives	Sales forecasting	Order planning/ forecasting	Order generation
Option A: Conventional order management	Retailer	Retailer	Retailer
Option B: Supplier-managed inventory	Retailer	Manufacturer	Manufacturer
Option C: Co-managed inventory	Retailer	Retailer	Manufacturer
Option D: Retail vendor-managed inventory	Retailer	Manufacturer	Manufacturer

Figure A.3 CPFR alternatives.

forecasts. But an alternative CPFR approach—called "Option A" in the list of CPFR alternatives shown in Exhibit 3 (Fig. A.3)—envisions the retailer as the lead partner, providing the enabling technology platform to generate sales and order forecasts that suppliers can use to improve order fulfillment. The retailer has primary responsibility for the reference forecast.

West Marine is an unabashed proponent of Option A. It was our choice because we believed it was more likely to deliver the improvements needed to achieve a scale that would influence enterprise-wide results. So we focused on getting our buyer-driven forecasts and processes in line before reaching out to suppliers. Our position has been that there needed to be a reference forecast at the outset, and it's the buyer who has to own that. Why spotlight the buyer-driven forecast? Two reasons: (1) the buyer usually drives the key events like promotions and assortment changes that crack the "bullwhip" in the supply chain, and (2) the buyer-driven forecast depends on only one technology platform and is therefore scalable across many items and suppliers with similarly accurate results.

However, our advanced replenishment system did not relieve us of the need to collaborate with suppliers with respect to changes in demand. Each supplier and product presents different opportunities and solutions. A supplier for a product featuring generic components that uses short-cycle final assembly or packaging processes may be able to respond to demand changes with reasonably short lead times without the need to expedite or without production disruptions. But a supplier of a product with individualized components or specialized packaging cannot respond as quickly. In the world of retail, longer lead-time products simply must be planned for and fully forecasted ahead of their sourcing and manufacturing lead times. For example, if the supplier's order cycle is six weeks and the demand changes occur inside two weeks, demand will be satisfied only if there's sufficient excess stock in the supply chain to cover the new demand.

West Marine has made a significant commitment to working with suppliers to match supply and demand. We have agreed together to the goal of planning demand accurately to accommodate their materials planning horizons. For instance, our normal lead time for planning promotional sales events is about ten weeks before the promotion begins. If the products promoted are of import origin, the supplier's lead time is significantly longer. With this in mind, most key promotional events are not only strategically planned but entered into the forecasting system months in advance to enable suppliers to fulfill them. Since this is a collaborative engagement, the supplier is a full participant in supporting and approving the forecast, which they receive weekly. West Marine also measures our forecast accuracy in addition to suppliers' shipping performance and other metrics. To make our commitment very clear to skeptical suppliers, we have also guaranteed our forecasted purchases. Generally, our suppliers believe in and act upon this guarantee.

West Marine supports its CPFR program by providing the following reports to its collaborative supplier teams: (1) a weekly automated e-mail containing our current purchase forecast by item and week for a year, (2) a weekly automated e-mail containing key in-stock and late-shipping information, covering the metrics achieved and the items that failed to meet goals, and (3) a monthly automated e-mail giving audit-level detail and performance measurement on each supplier's shipping record based on the supplier's advanced ship notices.

Multiple Routes to Supplier Buy-In

West Marine's supplier education program is the centerpiece of our strategy to improve our performance. Prior to our CPFR program, the only contact a supplier might have with West Marine (outside of the traditional buyer–seller relationship) was usually with a customer-service clerk.

Today, all of our collaborative suppliers join our monthly and quarterly meetings. Before a supplier can enter a collaborative engagement with us, one of its senior executives must sponsor the arrangement, and the supplier must name a point person to partner with our "supply-chain captain." We encourage the point people on both sides to be powerful collaboration champions in their respective organizations. Some suppliers have appointed Six Sigma Master Black Belts as their point people—a real vote of confidence in the program. At least 10 suppliers have dedicated vendor associates located at West Marine's facilities.

Our CPFR program demands that West Marine and its suppliers address supply chain performance as a cross-functional business process. Fundamentally, CPFR has caused West Marine to adopt cross-functional management and accountability, and it is producing similar outcomes for our suppliers. As they adopt team approaches to conducting supply chain activities with us, they also start to harmonize conflicting "silo" goals and metrics throughout their operations.

We have several forums for our CPFR engagements with suppliers. An engagement usually begins with a quarterly meeting. Since CPFR aims at holistic, long-term solutions, participation in a quarterly kick-off session requires us and the supplier to bring cross-functional teams who all contribute to the results and who are important stakeholders in the decisions that will be made. These stakeholders must include senior sales, marketing, and merchandising personnel along with the decision makers who manage the relationships for buying, forecasting, inventory control, production planning, distribution, and transportation.

The quarterly meetings are held at our offices or at the supplier's location; they usually entail a half day devoted to supply chain collaboration and a half day devoted to sales and marketing planning. The supply chain portion will include a session on West Marine's bottom-up forecasting process and a wide-ranging discussion of initiatives and opportunities. The initiatives are detailed by our supply chain captains, who must manage the solution process and timeline. Since senior executives are present, each side is expected to commit to resources and timelines for the improvements identified. We don't advocate a formal collaboration agreement, but we do expect a flexible commitment to our mutual success from both sides. (In our experience, bringing the lawyers into a contractual review often creates an insurmountable hurdle.)

As with the quarterly meetings, the "routine" or monthly meetings have defined formats and expectations. The format is usually a conference call attended at least by the two partners—our supply chain captain and theirs. The meeting is usually kept to about 30 minutes to make it efficient and effective and to encourage other key participants to attend, since they can expect the meeting will be concise, businesslike, and productive. (The president of one major supplier has not missed a routine meeting in over two years.) The agenda is three-part: review results (performance metrics for both sides); report on and manage current (or newly defined) initiatives by assigning clear owners, accountabilities, and deliverable dates; and resolve any supply chain constraints identified through review of the order forecast.

A special feature of our program is the twice-yearly Supply Chain Summit. The idea came from our category managers (our buyers). Once they learned about the structure of our collaboration effort and understood that we were committed to helping suppliers deliver the goods for their sales and marketing programs, they asked how we could "jump start" the program with more suppliers as soon as possible.

The summit conference follows a three-day format at a West Marine facility. It includes as many members of the collaborative teams as possible, and suppliers' senior sponsors and sales and marketing associates are required to attend.

The first and third days are reserved for quarterly meetings with about 25 suppliers. The second day is for plenary sessions designed to inform and support team-building and cultural change. We typically begin this day at our nearby distribution facility where members of each supplier team and their West Marine counterparts roll up their sleeves and physically unload and receive one of their shipments at our loading dock. Our DC team then leads a presentation on our logistics compliance program. Our CEO or general merchandise manager usually kicks off the afternoon to signal the extent of our commitment to collaboration. Other sessions address transportation management, electronic data interchange (EDI), and supply chain planning. The summit closes with a dinner that celebrates progress and partnership; attendees are invited to recognize others whose collaboration has helped them to break through silos or other barriers to improve performance.

There's one other collaboration forum that's important. We have initiated training sessions for key suppliers' top executives to develop insight and commitment to achieving what we call "breakthrough" performance. West Marine has picked 19 suppliers to participate. At the debut Breakthrough Session—a two-day retreat that was professionally facilitated—West Marine and the supplier partners signed a broad and powerful statement that reads: "We commit to breakthrough results for us and West Marine." Each executive agreed to commit his or her organization to the agreement. We did not submit this statement to legal review but did agree it would be placed prominently in our workplaces. At our facilities, the statement has been posted in every conference room where we meet with suppliers.

The response to these executive sessions has been astonishing: Our suppliers requested breakthrough training for all members of the collaborative engagement. We agreed, and we've now held a session for more than 100 of the breakthrough suppliers' associates.

Internal Collaboration Also Key

It has also been important to ensure that we are collaborating effectively inside our own four walls.

That's been particularly true within the merchandising function. Traditional retail merchandising organizations have two primary organizational "silos": buying and replenishment. For years, we had co-located our buying and replenishment functions and organized them as teams around product areas. But many silo viewpoints and practices remained, impeding holistic decision making and execution. Category managers and their assistants operated primarily as sole practitioners. Merchandise planners and replenishment analysts often dealt with supply chain management as a downstream effort. Communication between the co-located teams was often pretty poor. Merchandising associates owned sales and margin. Replenishment associates were responsible for inventory and service levels. Yet many of the specific responsibilities were ill-defined and inconsistent. For instance, since initial product forecasts and store assortment decisions were the purview of category management, some of the most important determinants of inventory performance were not the job of the inventory management group.

We have now transformed our merchandising organization into a best-practice model in which the perspectives of replenishment and merchandising are equally valued and balanced. Associates are expected to benefit from significant cross-training and career paths that incorporate both replenishment and merchandising. The category managers have given up tasks such as forecasting and store assortments that we consider more appropriate for planning and replenishment associates. Assistant category managers are no longer junior buyers. They are responsible for administrating product hierarchies, maintaining product information, and managing trade funds. The merchandise planner is now the business planner, responsible for financial plans and forecasts, key item forecasts, store assortments, lifecycle management, and promotional management and analysis. Meanwhile, our replenishment analysts manage the supplier collaboration process, purchasing, and store replenishment.

Another key to our performance improvement has been the close working relationship between logistics and planning and replenishment. Logistics chief Pat Murphy and I are kindred spirits in understanding the workings of the supply chain. We both recognized that merchandise planning determines the timing and volume of logistics activities. Early on, we established quarterly planning meetings and a weekly combined staff meeting to develop strategies and provide root-cause solutions. Of course, these groups had a common resource in the rich forecast information coming from planning and replenishment. They used this information to plan strategically and tactically together. Among the key benefits obtained were more standardized packaging for outbound shipments, which affected over 80 percent of the DCs' variable costs. Our use of standard packs has gone up from 15 percent to 55 percent.

Results Speak for Themselves

It has been a long, hard slog to do what we have done to date. From piloting our CPFR-based program with a handful of suppliers, we have now extended it to 200 suppliers and more than 20,000 items, representing more than 90 percent of our procurement spend. More than 70 of our top suppliers are loading the West Marine order forecast directly into their production planning systems. In-stock rates at our stores have come close to our goal of 96 percent in every store every week—even during our peak season. Forecast accuracy has risen to approximately 85 percent, and on-time shipments are now consistently better than 80 percent.

Our collaborative supply chain journey is by no means over. The competitive landscape will change, our supply base will shift, and our customers' demands will be different a decade from now. But we are pleased with where we are so far. The results, we believe, speak for themselves.

Store Operations Management: Making Lower Prices and Enhanced Customer Service Work Together to Drive Market Share at Giant Eagle, Inc.*

Giant Eagle, Inc. and H. B. Maynard and Company, Inc. team to implement a successful enterprise workforce management solution.

The Situation

Today, the business of retailing food is going through a dramatic, if not revolutionary, change.

- Competition has never been more vigorous with more than a dozen types of retailers vying for market share. Food retailers today include conventional supermarkets, super stores, super centers, membership clubs, combination (food and drug) stores, natural and organic outlets, limited assortment stores, convenience stores, and independent or chain fuel stations.
- The competition includes restaurants where dual-income couples, Generation X and the echo boomers, are fueling steady sales growth in the "food away from home" category.
- Rigorous competition—despite increased consolidation among food retailers—helped to contain food inflation to a yearly average of 2.5 percent from 1995 to 2004. Moreover, the cost of food continues to decline as a portion of family income—from 50 percent in the 19th century to 10.1 percent today.

The Challenge

At Giant Eagle, the Pittsburgh, Pennsylvania-based supermarket retailer [with] approximately 220 stores, its retail operations group has been able to boost the company's competitive edge by improving the work environment, work methods, work flows, standards, and labor management practices. The result has been improved workforce performance and reduced waste in the supply chain. For consumers, the program has meant enhanced customer service and reduced prices that were generated through the company's cost savings efforts.

**Source:* Maynard, "Giant Eagle Case Study," http://www.hbmaynard.com/CaseStudies/ArchivedCaseStudies.asp.

"We're facing competition everywhere, from mass merchants to dollar stores, from convenience stores to traditional grocers," said Giant Eagle Vice President of Retail Planning and Engineering, Dave Redick. "We're always looking for opportunities that will enable us to remain the market leader by improving customer service, reducing costs and providing better value for our customers." For nearly 75 years, Giant Eagle has built a reputation for excellent service, product quality and selection, and competitive prices that has helped them become the leading food retailer in western Pennsylvania and northeast Ohio.

The Solution

Giant Eagle asked Maynard to apply disciplines from Maynard's Workforce Performance Model™ to Giant Eagle's store operations to organize the project, analyze work flow, examine work methods, and set standards to conduct store efficiency initiatives, improve performance, and eliminate waste. "Our objective was to achieve sustainable improvements in Giant Eagle's cost structure while maintaining excellent customer service levels and while having the right people in the right roles," said Redick. "With Maynard's help, we are succeeding in doing this. Our plan is to take the savings from our efficiency initiatives and plow them back into improving the customer value proposition by lowering our everyday prices.

"With detailed workflow processes, our employees become more efficient, thereby increasing the time available by store employees for valuable interactions with customers." In addition, Giant Eagle looked to provide:

- A safer, cleaner, and more consistent work environment.
- A culture where store employees take pride in their work and their workplace.
- Improved data and tools to proactively manage performance.
- Visual tools and job aides to assist store employees in meeting expectations.

"Creating efficiencies like these improves our ability to remain price competitive in all of the markets where we operate," said Giant Eagle Executive Vice President and Chief Sales Officer Laura Karet.

Methodology

At the onset of the program, Maynard implemented the following process:

- Conducted an assessment to ensure that the program was properly aligned with Giant Eagle's business objectives.
- Implemented a pilot program at the store level within five major departments.
- Prepared Giant Eagle for a corporate store roll-out.
- Conducted a complete roll-out of Giant Eagle corporate and independently owned stores.

To address Giant Eagle's two main requirements—that the approach be comprehensive and that the results deliver real customer values—Maynard's solution entailed elements from its direct engineered process:

- Create a culture of discipline by introducing 5-S, a systematic approach for creating a clean and organized workplace.
- Improve work methods as a culture of discipline develops and associates begin to identify better ways to do their work.
- *Document work standards*—The improved work methods are documented and standard times are developed. Training tools are developed to communicate work methods and expectations to the work force.
- Staff based on new expectations.
- Hold people accountable across the board utilizing Best Methods, Productivity and Service levels.

The Results

The pilot program was successful in demonstrating the viability of the approach, and these stores have become models for others in the company.

The methodology developed is currently being rolled-out to all Giant Eagle stores with much success. The store personnel have embraced the culture change and are helping the company to drive waste out of their operations.

The cost efficiencies engineered into the program have allowed Giant Eagle the opportunity to twice lower everyday pricing on thousands of items across the store. Additionally, the project has resulted in a clean, more consistent and safer work environment, the ability to focus more time on the customer through the elimination of non-value added tasks and better allocation of personnel, providing the ability to put people in the right place at the right time to meet customer demand. "The end result is a more sophisticated system that is always up-to-date and more responsive to customer needs and the needs of employees," said Redick.

Redick said that the company is also witnessing a reduced turnover rate, providing better trained employees with longevity that customers eventually will come to know.

APPENDIX B

Lean Assessment Scorecard

	Internal Communication	Score (1–5)
1.	Management communicates with all levels of the organization on topics regarding organization goals and objectives at least twice per year.	
2.	Employees are able to accurately describe the organization's goals and how their jobs contribute to the achievement of those goals.	
3.	Employees receive feedback through a formal process concerning problems found in downstream processes or from the customer.	
4.	Management encourages Administrative, Operations, and Supply Chain & Logistics employees to work in groups to address performance, quality, or safety issues.	
5.	Employees at the operations level understand and use common performance metrics to monitor and improve the production processes.	
6.	Problems in the Administrative, Operations, and Supply Chain & Logistics processes are detected and investigated within 10 minutes of the first occurrence.	
7.	The concept of value stream mapping is understood, and all product families have been mapped and are physically segregated into the like process streams.	
	Internal Communication Category Score =	0%

	Visual Systems and Workplace Organization	Score (1–5)
1.	The distribution center, store (front-end and back-end), and office areas are generally clear of unnecessary materials, items, or scrap. Aisles are clear of obstructions.	
2.	The distribution center floor has lines that distinguish work areas, paths, and material handling aisles.	
3.	All employees are aware of good housekeeping practices, and operators consider daily cleanup and putaway activities as part of their job.	
4.	There is a place for everything, and everything is in its place. Every needed item, tool, material container, or part rack is labeled and easy to find.	
5.	Display boards containing job training, safety, operation measurables, production data, quality problems, and countermeasure information are readily visible at each production line or process and are updated continuously.	
6.	Check sheets describing and tracking the top quality defects are posted and are up-to-date at each workstation.	
	Visual Systems Category Score =	0%

	Employee Flexibility	Score (1–5)
1.	Employees are given formal training before doing a job on their own. Few defects or productivity-related slowdowns are attributable to new or inexperienced operators.	
2.	Product/Component travel distances have been measured, analyzed, and reduced by moving equipment and workstations closed together.	
3.	Equipment is "right-sized" for the operation/process. Employees have the ability to change speed to match the takt time. No "monuments" are present in the process.	
4.	Employees are cross-trained to perform other job functions, and operators work in at least two different jobs each day.	
5.	Processes and equipment are arranged to facilitate a continuous flow of work through the distribution center.	
6.	U-shaped cells have been designed and implemented to promote one-piece flow where appropriate (e.g., light assembly area).	
	Operator Flexibility Category Score =	0%

	Continuous Improvement	Score (1–5)
1.	There is a designated champion and a clearly communicated strategy for continuous improvement in the facility, with the necessary resources, organization, and infrastructure in place to support the process.	
2.	There is a formal suggestion process in place to solicit ideas for improvements from all employees and to recognize their participation.	
3.	Employees have been trained in continuous improvement methods and have been affected by or participated in continuous improvement events.	
4.	Employees know the eight wastes are actively involved in identifying wastes in their processes/areas, and are empowered to work to reduce and eliminate the wastes.	
5.	Continuous improvement kaizen projects/events are structured, planned, and implemented. Successes are recognized and expanded throughout the facility.	
6.	Most improvements made throughout the stores, distribution centers, and offices are made daily and involve little or no expense to implement.	
7.	Product/Process value streams undergo examination for continuous improvement on a regularly scheduled basis.	
	Continuous Improvement Category Score =	**0%**

	Mistake-Proofing	Score (1–5)
1.	Employees have been trained in the basics of mistake-proofing, and there is a team responsible for analyzing production defects and identifying mistake-proofing opportunities.	
2.	Mistake-proofing devices and methods have been implemented or are being developed to eliminate the top production defects for each work area in the plant.	
3.	Parts, products, and components have been analyzed to identify design opportunities to eliminate waste and improve productivity.	
4.	Employees are empowered to stop an activity when an error or defect is found or when they cannot complete their process according to the Standard Operating Procedure (SOP).	
5.	Manual processes or tasks have been equipped with mechanical checks to aid human judgment whenever possible.	
6.	Equipment and processes are equipped with call (andon) lights or signals that bring attention to situations requiring assistance with a problem or the replenishment of supplies.	
	Mistake-Proofing Category Score =	**0%**

	Quick Changeover/Setup Reduction	Score (1–5)
1.	Shift (opening, receiving, shipping, etc.) and wave startups are scheduled in advance and communicated to inform all workers that these events are on that day's schedule.	
2.	Shift and wave teams are in place, have received training on setup time reduction procedures, and are actively improving changeover methods.	
3.	Shift and wave startups are done frequently (wave only) and typically take less than 10 minutes.	
4.	Startup/Wave time is visibly tracked and posted at each workstation where work is performed.	
5.	Shift and wave startup procedures are standardized and repeated in other areas of the plant. Standard procedures and checklists are visible and followed	
6.	Special tools and equipment have been developed and implemented to reduce the time and labor involved in the startup process.	
	Quick Changeover Category Score =	**0%**

	Quality - Inbound, Outbound, and Administrative	Score (1–5)
1.	Zero defects from suppliers is a policy.	
2.	The company quality system is effectively implemented and compliant with a national standard such as ISO-9000.	
3.	FMEA (failure modes and effects analysis) is in place (feedback, root cause, etc.).	
4.	Material review board/discrepant material disposition is in place.	
5.	Supplier quality systems are in place.	
6.	Internal scrap loss is less than 1 percent of cost of goods sold.	
7.	Returned material to vendors is less than 0.1 percent of sales.	
	Quality I/B, O/B and Administrative Category Score =	**0%**

	Supply Chain	Score (1–5)
1.	Suppliers are involved in continuous improvement efforts with the company.	
2.	Performance to delivery policy (on-time) is better than 98 percent.	
3.	Quality performance of the suppliers exceeds 98 percent.	
4.	Electronic communication with suppliers is used to trigger release of supplies under a kanban system.	
5.	The company has regular input to the suppliers to improve design and performance characteristics of the supplied parts.	
6.	Cost reduction goals with suppliers are documented and tracked.	
7.	Sevice complaints with suppliers are resolved within 24 hours.	
	Supply Chain Category Score =	**0%**

	Balanced Flow	Score (1–5)
1.	There is an effort to level assembly, packaging, and administrative schedules by requiring suppliers to schedule frequent, smaller deliveries over the period.	
2.	Changeovers in assembly, packaging, and the office are made to support the concept of running to demand for all products, and not to support long production runs, WIP inventory buffers, daily short ship emergencies, etc.	
3.	Takt time is known by all associates and determines the pace of assembly and packaging in the facility.	
4.	Assembly and packaging are facilitated through value stream managers.	
5.	Processes on assembly and packaging lines or in cells (including the office) are balanced or leveled so the difference between cycle times of linked processes is negligible.	
6.	When demand volume changes, assembly, packaging, and administrative processes are rebalanced or redesigned to flex up or down the process cycle times to correspond to the new takt time.	
7.	When demand volume changes long term, supermarket and POUS levels are adjusted to meet the new takt time.	
	Balanced Flow Category Score =	0%

	Total Productive Maintenance	Score (1–5)
1.	Maintenance team managers and workers have been trained in the basics of TPM.	
2.	Machines and equipment have all necessary safety guards in place. Safety devices are in working order, and equipment is locked out immediately when broken down or when otherwise appropriate.	
3.	Preventive maintenance activity lists are posted in work areas, and item completions are tracked over time.	
4.	Accurate and visible maintenance records are kept up to date and posted nearby for all production and support equipment.	
5.	Preventive maintenance activities are focused on increasing process utilization and minimizing cycle time variation.	
6.	Preventive maintenance responsibilities are defined for both maintenance and production workers.	
7.	Time is allowed in the daily production schedule for workers to perform their preventive maintenance and cleaning duties.	
	Total Productive Maintenance Category Score =	**0%**

	Pull System (Store, Warehouse, and Office)	Score (1–5)
1.	Each assembly and packaging cell, line, or process has displayed, visually, the target and actual hourly output as well as the shift's production requirements and timing.	
2.	All office, assembly, and packaging managers and supervisors have been trained in the principles and implementation of material pull systems.	
3.	Material and information flow or movement in the office, assembly, and packaging areas are based on the make one move one concept, or are dependent on individual pull signals, via kanban, etc., from downstream workstations as information, parts, or materials are consumed.	
4.	Downstream processes are pulling material from upstream processes. Upstream production schedules are dependent on downstream use.	
5.	Office, packaging, and assembly lines/cells are capable of adapting to changes in customer demand by changing only one production schedule at the pacemaker process.	
6.	Office, packaging, and assembly supervisors are not motivated to produce more parts than the subsequent process requires.	
	Pull System Category Score =	**0%**

	Standardized Work	Score (1–5)
1.	Standard operating procedures have been developed for each process or cell and are used to train employees.	
2.	Every store, distribution center, and office process has its SOP posted within view of the worker performing the process.	
3.	The takt time (i.e., demand rate) for each product and/or service was used as the basis for the processing time for each operation and the process staffing requirements.	
4.	The process of job design and standardization involves operators as well as support personnel.	
5.	Frequently repeated, non-value-adding operations in the facility, such as setups, startups, quality checks, preventative maintenance, cleanup, etc., are visually standardized and updated.	
6.	Employees individually perform their processes according to the process sheets or SOP'S and make few method or technique errors. Any errors are recorded and tracked.	
	Standardized Work Category Score =	**0%**

	Engineering	Score (1–5)
1.	Engineering personnel are aware of, involved with, and trained in lean principles.	
2.	Systematic efforts are in place to reduce product variation and the number of items (part numbers) in the system.	
3.	Engineering has organized its activities along value streams.	
4.	Engineering processes are organized visually, and the workplace shows evidence of visual indicators to show the status of work.	
5.	Engineering processes have been balanced to create flow and reduce lead time within the engineering department.	
6.	Engineers routinely go to the location of a problem in production to assess the actual situation and communicate with the production operators to obtain their input.	
7.	Performance measures such as lead time and velocity are used to measure the department and establish goals for continuous improvement.	
	Engineering Category Score =	0%

	Performance Measurement	Score (1–5)
1.	Numerous and detailed financial reports have been replaced by a few key measures of enterprise performance.	
2.	Traditional cost accounting measures and individual/department efficiency measures have been replaced by value stream performance measures.	
3.	Performance results are communicated openly to all employees and are visually posted to show status and progress.	
4.	Employees understand how their individual efforts contribute to the overall results of the enterprise.	
5.	Individuals are rewarded for team-based performance rather than individual performance.	
	Performance Measurement Category Score =	0%

	Customer Communication	Score (1–5)
1.	There is a standard system in place for collecting customer satisfaction information and data.	
2.	Customer requirements are identified and communicated throughout the supply chain.	
3.	Customer complaints are handled the same day they are received in less than 2 hours.	
4.	Customers have regular and systematic input into the design and functionality of the products they buy.	
	Customer Communication Category Score =	**0%**

Lean Opportunity Summary and Graph

Category	Score
Internal Communication	0%
Visual Systems and Workplace Organization	0%
Operator Flexibility	0%
Continuous Improvement	0%
Mistake Proofing (Poka Yoke)	0%
SMED/Quick Changeover	0%
Quality	0%
Supply Chain	0%
Balanced Production (Assembly, Packaging & Office)	0%
Total Productive Maintenance	0%
Pull Systems (Store, Warehouse & Office)	0%
Standard Work	0%
Engineering	0%
Performance Measurement	0%
Customer Communication	0%

Enterprise Characteristic
Consider the overall average of all categories
0%–20% Traditional Administration, Operations and Supply Chain & Logistics
20%–40% Getting started with Lean
40%–60% Lean Progress
60%–80% Value Stream/Lean focus/integrated Supply Chain
80%–90% Lean Continuous Improvement Culture
>90% World Class Lean Retail and Wholesale

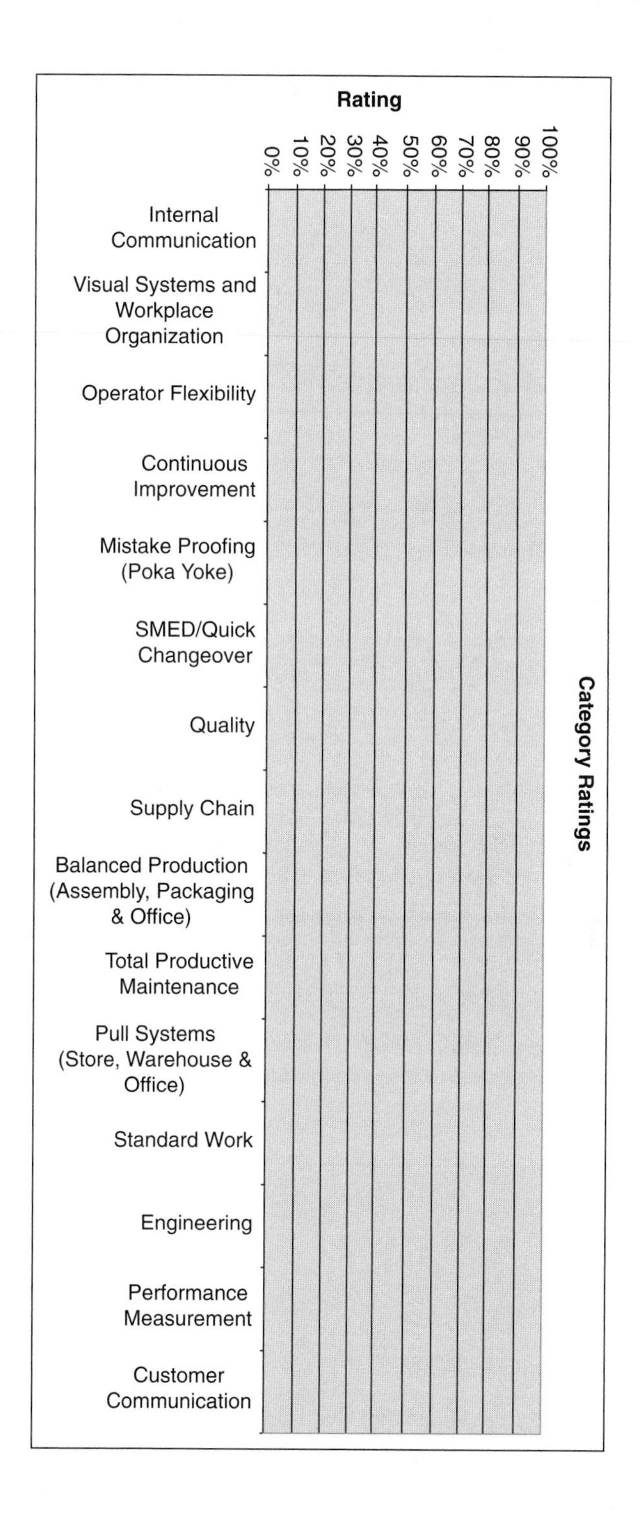
Rating
0%
10%
20%
30%
40%
50%
60%
70%
80%
90%
100%
Internal Communication
Visual Systems and Workplace Organization
Operator Flexibility
Continuous Improvement
Mistake Proofing (Poka Yoke)
SMED/Quick Changeover
Quality
Supply Chain
Balanced Production (Assembly, Packaging & Office)
Total Productive Maintenance
Pull Systems (Store, Warehouse & Office)
Standard Work
Engineering
Performance Measurement
Customer Communication
Category Ratings

REFERENCES

Chapter 1

Bowersox, Donald J., David J. Closs, and M. Bixby Cooper, *Supply Chain Logistics Management*, 4th ed. (New York: McGraw-Hill, 2013), pp. 139–140.

Bureau of Labor Statistics, "Industries at a Glance (Retail, Wholesale Trade)," www.bls.gov, accessed 2013.

Frohlich, Markham T., "Lessons from the Great Recession for Operations Management," *Journal of Operations Management*, Operations & Supply Management Forum, 2010, www.wpcarey.asu.edu/JOM, accessed 2013.

Guidon Performance Solutions, "Applying Lean Six Sigma Principles to 'Five Points of Transition' in a Retail Environment," www.guildonps.com, accessed 2013.

Jørgensen, Hans, Lawrence Owen, and Andreas Neus, "Making Change Work Study: Continuing the Enterprise of the Future Conversation," 2008, http://www-935.ibm.com/services/us/gbs/bus/pdf/gbe03100-usen-03-making-change-work.pdf, accessed 2013.

Murad, Andrea, "How the Great Recession Changed Generations' Money Views," April 26, 2012, www.FOXBusiness.com, accessed 2013.

Supply Chain Council, Supply Chain Operations Reference Model (SCOR) version 6.0, 2003, www.supply-chain.org, accessed 2013.

Chapter 2

Abernathy, Frederick H., John T. Dunlop, Janice H. Hammond, and David Weil, *A Stitch in Time: Lean Retailing and the Transformation of Manufacturing—Lessons from the Apparel and Textile Industries* (New York: Oxford University Press, 1999).

Berman, Barry, and Joel R. Evans, *Retail Management: A Strategic Approach*, 12th ed. (Upper Saddle River, NJ: Prentice Hall, 2013), pp. 23–24.

Levy, Michael, and Barton A. Weitz, *Retailing Management*, 8th ed. (New York: McGraw-Hill Irwin, 2012), pp. 6–8.

Porter, Michael, *Competitive Advantage: Creating and Sustaining Superior Performance* (New York: Free Press, 1998).

Smyyth LLC, "History of Retailing in North America," 2011, www.smyyth.com, accessed 2013.

Chapter 3

Harrison, Alan, Martin Christopher, and Remko I. Van Hoek, *Creating the Agile Supply Chain* (London: Institute of Logistics and Transport, 1999).
Myerson, Paul A., *Lean Supply Chain and Logistics Management* (New York: McGraw-Hill, 2012), pp. 11–15.
Womack, James P., Daniel T. Jones, and Daniel Roos, *The Machine That Changed the World* (New York: Free Press, reprinted 2007).

Chapter 4

Forrester Research, "US Online Retail Forecast," February 2011, www.forrester.com, accessed 2013.
Hawks, Karen, "What Is Reverse Logistics?," *Reverse Logistics Magazine*, Winter/Spring 2006, www.rlmagazine.com/edition01p12.php, accessed 2013.
Jargon, Julie, "Latest Starbucks Buzzword: 'Lean' Japanese Techniques," *Wall Street Journal*, August 4, 2009, p. A1.
J.P. Morgan Internet User Survey 2010, www.jpmorgan.com, accessed 2013.
Kaizen Institute Ltd., "Kaizen Leads the Way as Sonae MC Moves from Training to Innovation," www.kaizen.com, accessed 2013.
Myerson, Paul A., *Lean Supply Chain and Logistics Management* (New York: McGraw-Hill, 2012), p. 6.
Sheffi, Yossi, "The Value of CPFR," MIT Center for Transportation and Logistics, 2002, http://web.mit.edu, accessed 2011.
Traub, Todd, "Wal-Mart Used Technology to Become Supply Chain Leader," *Arkansas Business*, July 2, 2012, www.ArkansasBusiness.com, accessed 2013.
UPS, "UPS Reverse Logistics Results in a 75% Repeat Customer Rate," http://thenewlogistics.ups.com/customers/reverse-logistics/, accessed 2013.
Waller, Matt, Eric M. Johnson, and Tom Davis, "Vendor-Managed Inventory in the Retail Supply Chain," *Journal of Business Logistics* 20, no. 1 (1999).

Chapter 5

Anderson, Arnold, "Retail Layout Strategies," *Houston Chronicle*, http://smallbusiness.chron.com/retail-layout-strategies-11464.html, accessed 2013.
Menlo Worldwide Logistics, a division of Con-way, www.con-way.com, accessed 2011.
Myerson, Paul A., *Lean Supply Chain and Logistics Management*" (New York: McGraw-Hill, 2012), Chaps. 3 and 5.

Chapter 6

Burnsed, Brian, and Emily Thornton, "Six Sigma Makes a Comeback," *BusinessWeek*, September 10, 2009.
Myerson, Paul A., *Lean Supply Chain and Logistics Management*" (New York: McGraw-Hill, 2012), Chap. 6.

Kurt Salmon, "The New Supply Chain: Driving Top-Line Growth through Speed, Scope and Specialization," July 1, 2013, www.kurtsalmon.com, accessed 2013.

Chapter 7

Aberdeen Group, "Best-in-Class Marketers Drive Enhanced Customer Loyalty," *Analyst Insight*, August 2010, www.aberdeen.com, accessed 2013.

Alturos, "Lean in Retail," www.alturos.co.uk, accessed 2013.

Ansoff, Igor, "Strategies for Diversification," *Harvard Business Review* 30, no. 5 (1957).

Bacheldor, Beth, "POS Data Sharing—Could the New Merchant Customer Exchange Spur Retailers to Share More with Suppliers?," Best Practices ecbp Commerce, September 11, 2012, www.ec-bp.com, accessed 2013.

Bhargava, Rohit, "Twelve Big Ideas Transforming the World of Retail Now," presented at Shop.org Merchandising Summit, July 2012, www.slideshare.net, accessed 2013.

Bowersox, Donald J., David J. Closs, and M. Bixby Cooper, *Supply Chain Logistics Management*, 4th ed. (New York: McGraw-Hill, 2013), p. 396.

Bradley, Peter, "The Skinny on Lean," *DC Velocity*, 2006, www.dcvelocity.com, accessed 2011.

Council of Supply Chain Management Professionals, "Supply Chain Management Definition," www.cscmp.org, accessed 2013.

Curtis, Tony, Jin An, and Robert Gettys, "Solving Retail Problems Using Lean Six Sigma," Accenture white paper, 2008, www.accenture.com, accessed 2013.

Ferdows, Kasra, et al., "Rapid-Fire Fulfillment," *Harvard Business Review* 82, no. 11 (November 2004).

Heizer, Jay H., and Barry Render, *Operations Management*, 11th ed. (Boston: Pearson, 2013), p. 413.

Levy, Michael, and Barton A. Weitz, *Retailing Management*, 8th ed. (New York: McGraw-Hill Irwin, 2012), p. 221.

Myerson, Paul A., *Lean Supply Chain and Logistics Management* (New York: McGraw-Hill, 2012), Chap. 1.

Porter, Michael, *Competitive Advantage: Creating and Sustaining Superior Performance* (New York: Free Press, 1998).

Pitney-Bowes Software, "10 Common Mistakes in Retail Site Selection," www.pb .com/software, accessed 2013.

RetailSystems.com, "History of Retail Systems," www.retailsystems.com, accessed 2013.

Sassi, Cynthia, "How Retailers Can Use Social Media to Attract More Customers," *SmartBlog on Social Media*, May 21, 2013, www.smartblog.com, accessed 2013.

Sika Retail, "Retail Systems: A Primer," October 18, 2006, http://it.toolbox.com/blogs/ retail-erp-systems/retail-systems-a-primer-12358, accessed 2013.

Siwicki, Bill, "Retail Marketers Focus Mobile Investments on E-mail and Paid Search," *Internet Retailer*, July 16, 2013, www.internetretailer.com, accessed 2013.

Stevens, Ruth P., "CRM for Retailers," *1to1 Magazine*, August 18, 2003, www .ruthstevens.com, accessed 2013.

Swedberg, Claire, "Marks & Spencer to Tag Items at 120 Stores," *RFID Journal*, November 16, 2006, www.rfidjournal.com, accessed 2013.

Weiner, Jeff, "Future of Retail," a PSFK report, www.psfk.com/future-of-retail, accessed 2013 at www.slideshare.net.
Whole Foods, "Hiring Process," www.wholefoodsmarket.com, accessed 2013.
Xie, Ying, "Use of Information Technologies in Retail Supply Chain: Opportunities and Challenges," Abstract from POMS 20th Annual Conference, Orlando, Florida, May 1–4, 2009.

Chapter 8

Business Wire, "Starbucks Celebrates Five-Year Anniversary of My Starbucks Idea," March 29, 2013, www.businesswire.com, accessed 2013.
Ireland, Ronald, and Mary Adamy, "Migrating the Model," Oliver Wight Americas, Inc., March 2010, www.bus-ex.com, accessed 2013.
Levy, Michael, and Barton A. Weitz, *Retailing Management*, 8th ed. (New York: McGraw-Hill Irwin, 2012), pp. 308, 354–356.
Lukic, Radojko, "The Effects of Application of Lean Concept in Retail," *Economia Seria Management* 15, no. 1 (2012), pp. 88–98.
Myerson, Paul A., *Lean Supply Chain and Logistics Management* (New York: McGraw-Hill, 2012), Chap. 10, p. 109.
Oracle Retail Merchandising System, "Optimize Your Operations," data sheet, www.oracle.com/, accessed 2013.
Parker Avery Group, "Inventory Planning Methods: The Proper Approach to Inventory Planning," www.parkeravery.com, accessed 2013.
Småros, Johanna, et al., "Retailer Views on Forecasting Collaboration," Logistics Research Network Annual Conference, Dublin, Ireland, September 9–10, 2004.
Smith, Larry, "310—The Ultimate Retail Supply Chain: Connecting the Consumer to the Factory," presentation at GS1 Connect 2012, "Business Moving Forward," June 4–7 2012, www.gs1connectevent.org, accessed 2013.
Stein, Scott, "Kidtops: Best Buy to Sell Toshiba Satellite L635," CNET, September 23, 2010, www.cnet.com, accessed 2013.
Walmart, "Wal-Mart Introduces Exclusive New Line of GE Small Home Appliances," press release, May 31, 2006.
Waters, Shari, "How to Negotiate Prices and More: Bargaining Tips for Retailers," About.com Retailing, www.about.com, accessed 2013.

Chapter 9

Berman, Barry, and Joel R. Evans, *Retail Management: A Strategic Approach*, 12th ed. (Upper Saddle River, NJ: Prentice Hall, 2013), pp. 326–327.
Burns, Emily, et al., "Wegman's Bakery Freezer Reorganization," paper for Multidisciplinary Senior Design Conference, Rochester, NY, 2011.
Gleeson, Frank, and Alan Landy, "Doing More with Less: Lean Retailing," NACS show presentation, October 7, 2010, accessed 2013.

Guidon Performance Solutions, "Leveraging Lean Store Operations to Improve the Customer Experience," July 23, 2009, www.guidonps.com, www.slideshare.net/guidon/guidon-lean-retail-webinar-leveraging-lean-store-operations-to-improve-the-customer-experience, accessed 2013.

Grannis, Kathy, "National Retail Security Survey: Retail Shrinkage Totaled $34.5 Billion in 2011," National Retail Foundation, June 22, 2012, www.nrf.com, accessed 2013.

Heizer, Jay H., and Barry Render, *Operations Management*, 11th ed. (Boston: Pearson, 2013), pp. 13–19, 426.

Kaizen Institute Ltd., "Kaizen Leads the Way as Sonae MC Moves from Training to Innovation," www.kaizen.com, accessed 2013.

Kumar, Piyush, "The Competitive Impact of Service Process Improvement: Examining Customers' Waiting Experiences in Retail Markets," *Journal of Retailing* 81, no. 3 (2005), pp. 171–189.

Levy, Michael, and Barton A. Weitz, *Retailing Management*, 8th ed. (New York: McGraw-Hill Irwin, 2012), pp. 434, 508–509.

Pelger, Ron, "In the Trenches," *Produce News*, September 23, 2014, http://www.producenews.com/index.php/buyer-mnu/in-the-trenches/3126-2936, accessed 2013.

Purpura, Linda, "Efficient Consumer Response," *Supermarket News*, September 7, 1998, www.supermarketnews.com, accessed 2013.

Pyzdek, Thomas, "How One Retailer Competes Using Lean Six Sigma," Pyzdek Institute, www.sixsigmatraining.org, accessed 2013.

Raj, Devesh, "How Lean Services Lead to Lower Costs and Happier Customers," Interview with Devesh Raj, Knowledge@Wharton and Boston Consulting Group, http://knowledge.wharton.upenn.edu/papers/sponsor_collaborations/2013_02_BCG_Wharton_Lean_Services.pdf, accessed 2013.

Swedberg, Claire, "American Apparel Adopting RFID at Every Store," *RFID Journal*, February 8, 2012, www.rfidjournal.com, accessed 2013.

Tang, Angie, and Sarah Lim, *Retail Operations: How to Run Your Own Store*, 2nd ed. (Singapore: Pearson/Prentice Hall, 2008), pp. 137–145.

Chapter 10

Bradley, Peter, "The Skinny on Lean," *DC Velocity*, 2006, www.dcvelocity.com, accessed 2011.

Brenner, Mark, "Warehousing Gets Lean and Mean," *Labor Notes*, May 11, 2010, www.Labornotes.org, accessed 2013.

Enna.com, "Amazon Taps Kaizen to Deliver the Goods," February 15, 2013, http://www.enna.com/lean-articles/lean-outside-manufacturing/Amazon-taps-Kaizen-to-deliver-thegoods/, accessed 2013.

Forger, Gary, "Menlo Gets Lean," *Modern Materials Handling*, November 1, 2005, pp. 1–2.

Gaunt, Ken, "Are Your Warehouse Operations Lean?," *Universal Advisor* 3, no. 1–2 (2006).

Hawks, Karen, "What Is Reverse Logistics?," *Reverse Logistics Magazine*, Winter/Spring 2006, http://www.rlmagazine.com/edition01p12.php, accessed 2013.
McGladrey, "2010 Manufacturing and Wholesale Distribution National Survey," 2010, www.mcgladrey.com, accessed 2013.
Menlo Worldwide Logistics, a division of Con-way, "Lean Logistics," http://www.con-way.com, accessed 2011.
Myerson, Paul A., *Lean Supply Chain and Logistics Management* (New York: McGraw-Hill, 2012), Chap. 8.
Ryder Logistics, "Five Lean Guiding Principles," www.ryder.com, accessed 2011.

Chapter 11

None.

Chapter 12

Friesen, George, "Lean Drives Engagement and Engagement Drives Retail Sales," Workforce Solutions Group, St. Louis Community College, January 21, 2011, www.workforcesolutions.stlcc.edu/, accessed 2013.
Jørgensen, Hans, Lawrence Owen, and Andreas Neus, "Making Change Work Study: Continuing the Enterprise of the Future Conversation," 2008, www-935.ibm.com/services/us/gbs/bus/pdf/gbe03100-usen-03-making-change-work.pdf, accessed 2013.
Leahy, Sir Terry, "Every Little Helps: Tesco's CEO Talks Business," Interview with Sir Terry Leahy, October 7, 2010, www.londonbusinessforum.com, accessed 2013.
Rummler, Geary A., and Alan P. Brache, *Improving Performance*, 2nd ed. (San Francisco: Jossey-Bass, 1995), pp. 126–133.

Chapter 13

Ackerman, Alliston, and Alarice Padilla, "Black & Decker HHI Puts CPFR to Action," *Consumer Goods Technology*, October 20, 2009, http://consumergoods.edgl.com, accessed 2013.
Curtis, Tony, Jin An, and Robert Gettys, "Solving Retail Problems Using Lean Six Sigma," Accenture white paper, 2008, www.accenture.com, accessed 2013.
Deloitte, "The 2013 American Pantry Study," www.deloitte.com, accessed 2013.
DuPont, "The DuPont Production System and Asset Productivity Consulting," www.dupont.com, accessed 2013.
Faletski, Igor, "5 Paradoxes Shaping the Future of Mobile Commerce," February 16, 2012, www.mashable.com, accessed 2013.
GoLeanSixSigma.com, "Lean Six Sigma Success Stories in the Retail Industry: Staples Inc.," www.goleansixsigma.com/lean-six-sigma-success-stories-in-the-retail-industry, accessed 2013.
Guarraia, Peter, Gib Carey, Alistair Corbett, and Klaus Neuhaus, "Lean Six Sigma for the Services Industry," Bain Brief, Bain & Company, May 20, 2008, www.bain.com, accessed 2013.

Hawks, Karen, "What Is Reverse Logistics?," *Reverse Logistics Magazine*, Winter/Spring 2006, www.rlmagazine.com, accessed 2013.

HDA, "Inventory Impact of HDA's Supply Chain," Case Studies, www.hdainc.com, accessed 2013.

Kennedy, Kelly, "Five Big Data Trends Revolutionizing Retail," ZDNet, August 16, 2013, www.zdnet.com, accessed 2013.

Mahoney, Michael, "Running Lean and Mean to Survive in E-Business," *E-Commerce Times*, March 2, 2001, www.ecommercetimes.com, accessed 2013.

Miller, Craig, "CPFR Case Study: Liquor Control Board of Ontario," GS1, 2005, http://www.gs1ca.org/docs/CPFR%20Case%20Study%20-%20LCBO.pdf, accessed 2013.

Siwicki, Bill, "Office Depot: A Lean, Mean Shopping Machine," *Internet Retailer*, June 14, 2012, www.internetretailer.com, accessed 2013.

"The 10 Keys to Global Logistics Excellence," *Supply Chain Digest* white paper, 2006, www.scdigest.com, accessed 2011.

Tenhiälä, Antti, "Collaborative Planning, Forecasting and Replenishment in European Grocery Retail," Seminar in Industrial Management, Helsinki University of Technology, Department of Industrial Engineering and Management, November 26, 2003, http://legacy-tuta.hut.fi/logistics/publications/CPFRinEurope.pdf, accessed 2013.

Chapter 14

Fine, David, Maia A. Hansen, and Stefan Roggenhofer, "From Lean to Lasting: Making Operational Improvements Stick," *McKinsey Quarterly*, November 2008, www.mckinsey.com, accessed 2013.

Hanna, Julia, "How Mercadona Fixes Retail's 'Last 10 Yards' Problem," *Working Knowledge*, Harvard Business School, July 19, 2010, www.bhswk.hbs.edu, accessed 2013.

Hard Rock Cafe, "Cafe Training Program," www.hardrock.com, accessed 2013.

Noe, Raymond A., *Employee Training and Development* (New York: McGraw-Hill, 2002).

Chapter 15

Biggart, Timothy, Laurie Burney, Richard Flanagan, and J. William Harden, "Is a Balanced Scorecard Useful in a Competitive Retail Environment?," *Management Accounting Quarterly* 12, no. 1 (Fall 2010), www.imanet.org/, accessed 2013.

Klipfolio, "KPI Examples: Key Metrics and KPIs to Help You Stay Competitive," www.klipfolio.com, accessed 2013.

Thomas, George, "Using the Balanced Scorecard to Improve Retail Operations," Process Excellence Network, May 2011, www.processexcellencenetwork.com, accessed 2013.

Trends and Outliers (blog), "Big Data Analytics: High Stakes for Retailers," TIBCO Spotfire, December 4, 2012, www.spotfire.tibco.com, accessed 2013.

Chapter 16

Consumer Reports, "Wegmans, Trader Joe's, Publix & Fareway Top Consumer Reports Supermarket Ratings," April 3, 2012, www.consumerreports.org, accessed 2013.

Konzek, Lindsay, and Jenel Stulton-Holtmeier, "12 Trends in Wholesale Distribution in 2013," *Modern Distribution Management*, December 25, 2012, www.mdm.com, accessed 2013.

McTaggart, Jennifer, "The Golden Years," *Progressive Grocer*, May 2012, www.progressivegrocer.com, accessed 2013.

Reinvestment Fund, "Understanding the Grocery Industry," September 30, 2011, www.cdfifund.gov, accessed 2013.

Vanson Bourne and NetSuite, "The Retail & WD Trends Report 2012: Mobile, Social and Direct Sales for Growth," September 2012, www.netsuite.com, accessed 2013.

Welty, Scott, "Retail 2020: Seven Trends Impacting Brick & Mortar Retailers," *Chain Store Age*, September 2, 2013, www.chainstoreage.com, accessed 2013.

What's Next, "Top Trends in Retail, Shopping and Leisure," www.nowandnext.com, accessed 2013.

Appendix A

Maynard, "Bob's Stores Case Study: Getting Lean in the Retail Sector: Bob's Stores Makes It Happen," www.hpmaynard.com, accessed 2013.

Maynard, "Giant Eagle Case Study: Making Lower Prices and Enhanced Customer Service Work Together to Drive Market Share," www.hpmaynard.com, accessed 2013.

Smith, Larry, "West Marine: A CPFR Success Story," *Supply Chain Management Review*, March 1, 2006, accessed 2013.

Note: Portions of Chaps. 5, 6, 10, 11, 12, 13, and 15 were excerpted with permission from the book *Lean Supply Chain and Logistics Management*, by Paul Myerson, published by McGraw-Hill Professional, © 2012.

INDEX